Al Jazeera, Freedom of the Press, and Forecasting Humanitarian Emergencies

This book reveals how Al Jazeera and its news coverage became a force for change politically, socially and culturally in the Middle East in general and the Arab world in particular.

It explores pre-Al Jazeera and post-Al Jazeera representations of humanitarian crises and identifies a potentially significant partnership between the news organization and humanitarian actors. By tracing the evolution of the news network, the book sheds new light on how Al Jazeera effected change in the Global South. The research identifies a significant relationship between Al Jazeera's news coverage and the ability to forecast international humanitarian actions, politically and militarily. It also explores the potential for continued partnership between humanitarian actors and news organization to identify crises in their early stages. Last, the book examines the distinct, original lexicon developed by Al Jazeera for humanitarian affairs and shows how the network influenced international media stylebooks and changed humanitarian coverage on key global issues.

A compelling examination of Al Jazeera's news operation that will be of interest to students and scholars of media studies, political communication, journalism and news reporting, international politics and the media, and Arab media.

Yehia Ghanem is an Egyptian journalist, writer and commentator now based in Washington, DC. For almost 20 years, he worked as a war correspondent covering conflicts in Serbia, Croatia, Bosnia, Yemen, Afghanistan, Democratic Republic of Congo, Sierra Leone and Angola. He has won awards for his coverage of the Bosnian Civil War. He was the Senior Editor of the prestigious Egyptian newspaper *Al-Ahram*, and later headed the Dar-Al-Hilal media conglomerate as Chairman of the Board. He has appeared as a panelist and keynote speaker at national, regional and international seminars and was a Fellow in Residence from 2013 to 2014 at the City University of New York (CUNY).

Routledge Focus on Media and Humanitarian Action
Series editors: Robin Andersen and Purnaka L. de Silva

Humanitarianism is defined by assumptions that guide global solidarity, and posits that all peoples are part of the same humanity, no matter who they are, what they believe or where they live. These principles suggest that when media show the suffering of others, global publics respond in ways that facilitate disaster relief and help alleviate pain. But reactions to crises are also shaped by those who bear witness, tell the stories, share the data, and take the pictures of communities rocked by crises. Media content can also help those humanitarians who seek to address root causes of disasters, or it can serve to obscure the causes in many ways.

This series explores the multiple intersections between media and the work of humanitarian actors, and offers critical analysis of media, its uses, its coverage, how it has changed, and how it is misused in the representation of humanitarianism. Authors identify cutting-edge uses of new media technologies, including big data and virtual reality, and assess the conventions of older legacy media. For movements toward global peace, all peoples should be represented at the table and have their voices heard, including those outside the media spotlight.

www.routledge.com/Routledge-Focus-on-Media-and-Humanitarian-Action/book-series/RFMHA

Media, Central American Refugees, and the U.S. Border Crisis
Security Discourses, Immigrant Demonization, and the Perpetuation of Violence
Robin Andersen and Adrian Bergmann

Al Jazeera, Freedom of the Press, and Forecasting Humanitarian Emergencies
Yehia Ghanem

Al Jazeera, Freedom of the Press, and Forecasting Humanitarian Emergencies

Yehia Ghanem

NEW YORK AND LONDON

First published 2021
by Routledge
52 Vanderbilt Avenue, New York, NY 10017

and by Routledge
2 Park Square, Milton Park, Abingdon, Oxon, OX14 4RN

Routledge is an imprint of the Taylor & Francis Group, an informa business

© 2021 Yehia Ghanem

The right of Yehia Ghanem to be identified as author of this work has been asserted by him in accordance with sections 77 and 78 of the Copyright, Designs and Patents Act 1988.

All rights reserved. No part of this book may be reprinted or reproduced or utilised in any form or by any electronic, mechanical, or other means, now known or hereafter invented, including photocopying and recording, or in any information storage or retrieval system, without permission in writing from the publishers.

Trademark notice: Product or corporate names may be trademarks or registered trademarks, and are used only for identification and explanation without intent to infringe.

Library of Congress Cataloging-in-Publication Data
A catalog record for this book has been requested

ISBN: 978-0-367-51573-7 (hbk)
ISBN: 978-0-367-51575-1 (pbk)
ISBN: 978-1-003-05450-4 (ebk)

Typeset in Times New Roman
by Apex Covantage, LLC

For all the journalists who languish in prisons around the world, you will be remembered and honored by the enemies you have made.

Contents

Acknowledgments		viii
Foreword by Robin Andersen		x
Preface		xii
	Introduction	1
1	A Voice in the Wilderness: The Early Years of Building Al Jazeera From the Ground Up	6
2	From South to North: Reversing the Flow of Information While Covering War and Disaster	21
3	A New Kind of Humanitarian Journalism: Partnerships, Coalitions, Research and Investigations	35
4	Case Studies and Al Jazeera's Next Phase: Protecting Journalists and Human Rights and Predicting Disaster	53
5	The Power of Words: Between Al Jazeera's Humanitarian Stylebook and the Hateful Rhetoric of Radio Rwanda	64
6	Ethics and Values of Good Journalism in a Dictatorial Environment	83
	Conclusion: Global Press Freedoms Under Attack	94
	Index	105

Acknowledgments

This book would never have been possible if not for my numerous and generous family and friends who have contributed to this book and, more than that, have enriched my life.

The professional journalists who kindly devoted so much of their precious time sitting patiently for hours answering my countless questions did so over a five-week period, for which I am eternally grateful. Their words and stories, their experiences and analyses make up the core of this work. I am indebted to Salah Negm, News Director of Al Jazeera International; Wadah Khanfar, former Director General of Al Jazeera Network; Ayman Gaballah, Managing Director of Al Jazeera Mubashir; Imad Musa, former head of Al Jazeera English Online; Sami Al-Haj, Director of Al Jazeera Public Liberties & Human Rights; Dr. Mohamed Mukhtar Al-Khalil, Director of Al Jazeera Center for Studies; and Nawaf Al-Mansouri, Manager of the Al Jazeera Network Creative Services.

To Robin Andersen, who has worked tirelessly over a period of months to shape this work in a way that would bring it to the light of day; to her I owe so much. To P.L. de Silva, who first recognized the value of my publishing on the topic of Al Jazeera.

To my dear friend Ann Schroeder, who has been a great supporter. I must also recognize the role of Caroline Sasseen, who was among the first to read the initial draft of the book and who offered insights and careful comments that were incredibly helpful.

To my prestigious *Al-Ahram* media conglomerate that afforded me the opportunity to enhance my journalistic skills from a novice journalist, to a war and political correspondent, to becoming a managing editor before being nominated and appointed as Chairman of the Board of one of the most prestigious and oldest media houses in the Middle East and Africa. To many *Al-Ahram* professionals, especially Ibrahim Nafea, former Chairman of the Board, I owe a great deal.

Acknowledgments ix

To my late mother, who insisted since my childhood that I should grow up to be a journalist, and my late father, who taught me that the WORD is the dividing line between heaven and hell; I will always cherish their memories and be proud of their influences, not only for bringing me to life, but more importantly, for making me human.

Finally, I thank my wife and children, who have suffered greatly over the years because of the ordeals I have gone through, especially of being accused, charged, sentenced and then exiled because of a trumped-up, politically driven case in Egypt. Their endurance, resilience and support have been the lifeline that helped me survive.

I hope this work will assist journalists, media professionals, media researchers, humanitarians and possibly even governments in realizing the imperative of freedom of expression, which at its heart is nothing less than freedom of the press.

Yehia Ghanem
Washington, DC
March 2020

Foreword

When I was first introduced to the topic of this book, I knew it would be a valuable contribution to our series. It is a study of the Qatar-based media channel Al Jazeera and how it has built an impressive network of professional reporters, news bureaus, institutional structures, partnerships with nongovernmental organizations (NGOs) and coalitions with other civil society groups so that it can now hit the ground reporting even before a humanitarian crisis is full blown. In the pages of this book, the author argues that experienced professional journalists, supported by a remarkable global news network, have the skills and ability to evaluate conditions and predict a crisis before it occurs. These abilities are the result of the years Al Jazeera has worked to develop an agile news outlet dedicated to telling the stories of the people of the Middle East. The network can now act as a partner, helping to augment relief efforts by alerting aid agencies to impending disasters. Early detection has become a strategy that many NGOs and government agencies have recognized as the next step in humanitarian action and an essential factor in saving lives. Indeed, the book covers this timely topic, and its contents are of immense value to humanitarian actors and scholars alike. But because of the powerful voice of its author, Yehia Ghanem, the book is so much more than that.

Ghanem shares his insights and experiences as one of the most formidable journalists in the Middle East, one who has worked in the region for three decades. He provides the reader, especially the western reader, with a rare insider's view into all the complexities of developing professional journalism in a region without a strong tradition of press freedoms. He brings his own personal history to these pages and with it a unique perspective and analysis of what it means to be a reporter in a region where journalism is a much different enterprise than it is in the west.

As an Egyptian, working for the oldest and once most prestigious newspaper in the Middle East, he has watched as his beloved *Al-Ahram* has fallen from such great heights to be now little more than a mouthpiece for the current dictatorial regime. It is no wonder, then, that the passion for freedom of expression and the power behind the arguments that speak so eloquently about the necessities of a free press can be felt on every page of this volume.

Few reporters have left their field and turned such an analytical eye back onto the profession they once practiced. Indeed, few journalists have directed their expert observational talents, on-the-ground skills and interview techniques to engage in such an extensive and valuable research project. Ghanem might not have taken on this project if it weren't for his prosecution in Egypt and subsequent exile, on trumped-up charges for doing the work of journalism there. In a very real sense, we have benefitted from his suffering and the ordeal of being locked in a cage in an Egyptian courtroom for 19 months. We might have been deprived of his personal perspective and experiences, experiences that he himself has mined for the larger lessons they teach us all about freedom and justice.

For all these reasons, I am honored to present this book in our series, which also tells a story about the cruel injustice of being falsely accused and condemned. Its publication is a liberation from the shackles of soul-deadening tyranny and the dark world of censorship and silence.

Robin Andersen
Guadeloupe
March 5, 2020

Preface

In the year 1290 B.C., the Prophet Moses led his people out of Egypt, escaping the tyrant Pharaoh, who persecuted the Israelites. The Pharaoh, who belonged, in most historical studies, to the eighteenth dynasty, was so vengeful he decided to lead his army and chase the Prophet and his people. The situation was desperate, since the Israelites were eventually caught between the Red Sea and the tyrant's army. Only Moses and a few of his confidantes still believed that God had not yet given his final word. Soon after, God inspired the Prophet to strike the waters and part the sea to enable the Israelites to cross. Despite this miracle, and those that came before, the Pharaoh and his army kept up the chase, only to meet their doom when the roaring sea closed in on them. After that day, these words have been deeply etched in Egyptian culture: "For every Pharaoh, there is always a Moses," meaning that no matter how long a tyrant lives and how strong he gets, he will always be beaten by a good man. Since we no longer live in a time of miracles, the story of Moses is now the story of courageous journalism, armed with values that include accuracy, fairness and balance, the equivalent of the Ten Commandments. Journalism, especially in the Middle East, has become the modern day "Moses" against the forms of tyranny and disaster, both human-made and natural. Tyranny, in many cases, is the cause of human-made disasters that result in humanitarian crises such as wars and conflict and even natural disasters set in motion by human endevours that exploit, fracture and diminsh the natual world. By extension, independent journalism's mission now entails standing up to crises and whoever shapes them by explaining the forces that cause them and using communication skills for relief efforts that aid those in distress. Jounalism and its practitioners can also facilitate the forecasting of emergencies and help set in motion strategies able to save lives.

Many media entities across the world have tried to take after Moses; some have succeeded on different levels, especially in democracies;

many were crushed by dictatorships; and quite a few shined a light that helped topple tyrants. Al Jazeera is one of the few that made its own history as a beacon of light, revealing human suffering and calling for global actions. This book tells the story of the network and the history of how it developed and how it can continue to advance the global commitment to humanitarianism and the alleviation of suffering. It also challenges global news organizations to stand for freedom of the press, while also foregrounding the principles of equality and humanism, dignity and justice in an age of increasing tyranny across many parts of the world.

Introduction

This is a book that considers the Qatar-based media network Al Jazeera from a number of perspectives, from different sources of information, from applied theoretical knowledge and also from the long experience of journalism in the Middle East. In the pages of this book are pointed insights from the journalists who help found Al Jazeera and went on to create a media institution unlike anything that came before it in the Middle East. Al Jazeera represents a particular type of journalism, and we learn the process and practices of those who trained the journalists, institutionalized the professional standards of journalism and introduced a balanced dialogue in a region not known for openly challenging the power and perspectives of the ruling state. Al Jazeera became the voice of the voiceless, where people who had been excluded from participating in public dialogue for so long found an outlet that valued their experiences, opinions and lives. These journalists, news editors and program directors helped nurture a network and created an organization that has grown into a formidable global presence with significant influence coming from a region with increasing complexity and difficult struggles.

As a global news organization that has been dedicated to developing humanitarian perspectives and news discourses, with an ongoing commitment to working with humanitarian partners, this study of Al Jazeera has broken new ground in the discipline of media and humanitarian affairs. It is well known that news coverage that shows the human suffering of those in crisis can lead to public awareness and help mobilize global audiences and their governments to take action to alleviate that suffering. Over the course of the twentieth century and up to the present day, aid professionals and relief agencies have developed a symbiotic relationship with the journalists who cover complex emergencies. Nowhere are these partnerships more essential to news coverage and humanitarian reporting than in the newsrooms and study centers of the Al Jazeera network. The book follows the evolution of the network as it came to realize it could help with information about disasters and crises even before they became full-blown emergencies.

The book begins with a detailed discussion of the building of Al Jazeera from the ground up. Since its first two-hour daily broadcasts that began in 1996, the network's mission to bring global news to the Arab world meant it would have to create a new kind of journalism. The determination to break from the old models of official news, news that featured the opulent ceremonies of those in power and the pomp and directives of heads of state, would require a different focus and an alternative set of priorities. Training and skills were important, but most essential was the opening of a new attitude toward the role of the media in the region. Journalists were encouraged to challenge official pronouncements of the state and offer differing opinions and attitudes toward the power centers of often dictatorial governments. The early focus of the network was on the human-centered narratives that told stories rarely heard about the people almost never featured on TV screens in the region. Al Jazeera focused its cameras on the common person, and journalists went into the streets to feature those who push the carts and toil to make a living in the Middle East. In creating a new kind of journalist and a new kind of media structure, Al Jazeera succeeded and continues to evolve as a primary source for information in the Middle East.

For the next phase of Al Jazeera, the human perspective the network had developed was applied to the reporting of conflict, and it showed the human face of war. On the ground with those on the receiving end of air wars, it offered a view divergent from the news media of the west, which celebrated the "smart" bombs and "surgical" strikes of the powerful destructive force of weaponry that often resulted in significant civilian casualties. Al Jazeera became the voice and the face of a region that had previously only seen itself as a reflection of the western press. It became a source for news that had become so commercialized in the west that it could no longer cover the vast regions of the globe with bureaus that had closed and budgets that demanded short-term coverage. Al Jazeera became the source for news of all kinds from the Global South, particularly when it came to the restricted fields of battle and conflict. International news divisions now featured the Al Jazeera logo, including western media and those based in the United States. This perspective led to the evolution of the network into a renowned international news organization as it reversed the flow of information to become a source for news from the Global South and also for the North.

This book also examines the theoretical foundations and practical structures that have allowed Al Jazeera to report the plight of people and communities caught in humanitarian emergences of all descriptions.

Once Al Jazeera news operations sent camera crews and reporters into humanitarian emergencies, those crises could no longer be ignored or, as in some cases, covered up. The genesis of Al Jazeera's news operation is detailed as well, as it developed news bureaus, staff and international liaisons to cover humanitarian emergencies. This endeavor is also an exploration of the sinews that enable Al Jazeera's structure and history to lead in this direction. Al Jazeera's commitment to covering all aspects of humanitarian actions has allowed it to be on the ground during crises well before any of its competitors. Equally important is the fact that reporters often remain in the field when others have left. Because of this commitment and the network's long history of humanitarian reporting over the years, the network has developed into a leading light, reporting on crises and humanitarian actors and their goals, practices, successes and failures. In the pre-Al Jazeera Arab world, without journalists documenting natural disasters and conflict, violence and their humanitarian consequences, reporters could not bear witness to humanitarian crises, and global publics were not informed about the human suffering in the region.

This study lays out how the network developed coalitions and partnerships that help augment news reporting of corruption and scandal, as well as coverage of complex emergencies. The history and genesis of the network's Center for Studies is here, along with detailed explanations and case studies of its successes in helping explain our world. At the center, experts and policy analysts provide background and context for news topics that serve to help news divisions find a balance between simply reporting the facts and finding narrative contexts that explain global events.

Media and Humanitarian Action: From Reporting to Early Detection

This book discusses the intersectionality between media organizations, humanitarian actors and the global emergencies that people, communities and countries face. As a result of covering crises, experienced reporters often develop the skills necessary to comprehend a looming crisis on the ground before it happens. Humanitarian journalists know how to ask questions and observe social and political forces, using the investigative and narrative skills to communicate the genesis and nature of humanitarian crises to others. These abilities become more important when they have editorial and institutional support, including the column inches, broadcast time and visual platforms at hand. From first responders to development organizations, and from urgent need to follow-up actions, liaisons with humanitarian partners have the potential to augment existing channels of early detection and to inform first responders in ways that could help save lives.

Chapter 5 introduces the network's stylebook, with in-depth interviews that explain the editorial practices that have resulted in an alternative lexicon for humanitarian reporting. Because words help to define our world and shape public attitudes toward disasters and their victims, editorial discussions of the terms and phrases used in reporting helps journalists and editors communicate to the public in a different way. With its in-depth interviews, this book gives readers an insider's view of the editorial discussions of words and their political implications and connotations. Al Jazeera looks and sounds so distinct from other news outlets because it has developed an alternative stylebook that refuses to call refugees "illegal immigrants" and finds ways to document the loss of civilian life in war without using such euphemisms as "collateral damage." Narrative, visual language and sound, including songs, take the entire audio-visual landscape of coverage into account. Language is an essential component of journalism motivated by the desire to aid those who suffer and find resolutions to conflict. This often means establishing a new lexicon and finding new words, words that are fundamentally about healing and uniting the peoples of the world, not dividing groups, ethnicities and those with differing belief systems. These practices stand in striking contrast to a lexicon of political division, conflict, belligerencies and war.

In this book I draw on my years of experience as a journalist, war correspondent, news editor and manager to lay out a set of practices essential for professional, ethical and effective reporting, especially under conditions where press freedoms are curtailed or restricted. I detail standards and ethical positions that help protect lives—especially in war zones. The need to secure and protect sources and the ways to prevent retaliation are explored, all the while looking to the objectives of journalism to inform the public and keep the powerful accountable by telling the human-centered stories that matter. Through these discussions, the standards that Al Jazeera brings to the news are compared with media that do little more than follow the dictates of the centers of power.

Push Back: Al Jazeera and Press Freedom Under Attack

Despite the significant global role Al Jazeera has played as a news innovator and its long and distinguished role in disseminating humanitarian information in the Arab world, the network has faced massive attacks by autocratic and tyrannical regimes in the region. Al Jazeera has come under attack and is now threatened and in danger of being silenced. As powerful regimes in the region are waging a fierce campaign against the network and the country from which it broadcasts, Qatar is under siege. In a push-back to freedom of the press in a region where people languished for decades in cultural prisons, massive attacks on the network escalated during the siege of Qatar that began on June 5, 2017. Saudi Arabia, the United Arab Emirates, Bahrain and Egypt presented an ultimatum: Qatar must shut down Al Jazeera. In the meantime, the western states and western media have not made a forceful stand in support of the network. As the region remains embroiled in conflict, the west has not lived up to the principles it has long stood for, including, first and foremost, humanitarianism and freedom of the press.

Today the void caused by the growing absence of independent media in the Middle East, and freedom of expression, is being filled by extremism. As the avenues of legitimate news and information are blocked, those left in the dark turn to xenophobia and sectarianism: the toxic consequences of tyranny. The full-blown influences of extremist groups such as Al-Qaeda and ISIS in Iraq and Syria are, at least in part, consequences of this type of repression.

If global media do not stand up to support each other, the world will remain in the dark, without a bright light to shine on the wrongs done to humanity. As conflict and violence continue to spread globally, including in the Middle East, global publics and humanitarian actors require more information, not less. Shuttering Al Jazeera and shutting down its ability to inform will only lead to more suffering.

With its commitment to the global publics that it serves and its dedication to independent journalism, guided by a stylebook shaped by humanitarian values and knowledge-based reasoning, Al Jazeera has developed from a single channel broadcasting for only two hours a day into a world media giant. Its global prestige has grown with its brand, and the network is recognized among the top ten worldwide. Independent media may not gain government support, but it guarantees public respect, an extended life and the potential influence to change the world for the better.

1 A Voice in the Wilderness
The Early Years of Building Al Jazeera From the Ground Up

Al-Ahram

In December of 1875, two Lebanese immigrants founded the *Al-Ahram* newspaper in Alexandria, Egypt. Over the past 143 years of continuous daily publications, *Al-Ahram* has grown to be one of the oldest media conglomerates and also a highly respected economic empire. I was one of the lucky few to join *Al-Ahram* in the late 1980s. Within a few years, I became the paper's war correspondent, trekking from one war zone to another across three continents. Since the late 1980s and until 1996, when Al Jazeera was born, I always felt that *Al-Ahram* was the most prestigious newspaper in the Middle East and Africa. It seemed to me that it was the lone voice in the wilderness crying out against humanitarian catastrophes, be they human-made conflicts or natural disasters. Indeed, we were in the middle of all wars, from the first Chechen one; to the Serbo-Croatian and Bosnian wars; to the war in Afghanistan between the Taliban and Northern Alliance to the wars in Congo, Sierra Leone and Angola, although for the sake of the truth, *Al-Ahram*'s war coverage was almost absent after October 6, 1973, and the war between Egypt and Israel, not to reappear until the end of the 1980s.

Since the end of the 1980s, *Al-Ahram*'s war coverage was penned by one war correspondent, myself. I was the only reporter covering conflict at the newspaper from the late 1980s until 2010. Even though war was my "beat," during my early years on the job, I hardly knew the crucial role media could play in directing and redirecting humanitarian relief to disaster-stricken areas. I must admit that I came to realize media's critical role only after I received a note from the Supreme Committee of Humanitarian Relief in Egypt at the end of 1994, informing me that I was to be decorated their "Man of The Year." The award came in recognition of my efforts to raise public awareness of the horrors and agonies of the Bosnians who were suffering. At that time, satellite television channels were newborns in the west and entirely unknown in the Middle East.

I received the award at the pinnacle of *Al-Ahram*'s success: we had a daily circulation of 1.9 million, including the international edition, 1.7 million for the weekly Friday edition, and readership was up fivefold. *Al-Ahram* was at the height of its influence, setting the agenda for local television and to a certain extent for many TV stations in the Arab world. In the second half of the 1990s, my war coverage was syndicated to at least 20 leading newspapers across the Arab world, which multiplied the paper's impact. It is worth mentioning that *Al-Ahram* was one of the few Arab media outlets at the time that covered wars in the Middle East.

Al-Ahram's prominence in conflict coverage remained almost unchallenged until Al Jazeera was born in 1996. After that, I started seeing colleagues from the channel in the same conflict zones that I covered. During those years, I saw that war reporters from Al Jazeera would be essential to the programming and that the channel would focus on covering conflicts more than other news topics. However, Al Jazeera's coverage of both human-made and natural disasters came to distinction later, at the beginning of the new millennium, and the reasons for this shift had to do with the unique phases of development the channel went through. It quickly grew to be a giant network in the region. Although we must give credit to *Al-Ahram* for keeping the tradition of war coverage in the news by having a sole war correspondent from the late 1980s up until 2010, it pales in comparison to the role Al Jazeera would play in crisis coverage ever since. The difference between *Al-Ahram*'s effort and that of Al Jazeera is massive, and that difference can be attributed to the organizational structures built by the network and the way it institutionalized war coverage. For this reason, *Al-Ahram*'s effort was almost *a cry in the wilderness*, whereas Al Jazeera's was *the antithesis* of *vox clamantis in deserto*.

The Early Years From 1996 to 2001: "The Opinion and the Other Opinion"

Until 1996 and for at least half a century, the Arab world and the Middle East, in general, had a media structure that could only be characterized as a monolithic block repeating monotonous speech, an industry controlled by dictatorial regimes that used media to consolidate their powers. Even in the 1990s when they had to allow private media, tyrannical regimes controlled them by controlling the businessmen who owned the media outlets. The main target of Al Jazeera since the first day of operation has been to break the monopoly maintained by dictatorships in the Middle East for so long. The way the new channel achieved this was by introducing a vanguard motto that heralded a new era in the region, not only in media but also, and more importantly, in political life. The motto was "Introducing the Opinion and the Other Opinion," in a region where one and only one opinion had been allowed for so long, that of the regimes. This change was a real revolution that profoundly impacted the heavily censored media. More importantly, it shook up many dictatorial regimes and pushed them to effect some positive changes.

During the years from 1996 to 2001, the perspective of providing a platform for a wide range of voices in the Middle East became a confirmed faith and a way to gain growing trust and credibility in the eyes of the people of the region. It also evoked extreme animosity toward Al Jazeera, and by extension Qatar, from almost all of the regimes in the area.

In his book, *Al Jazeera English: Global News in a Changing World*, Seib Philip argues that when Al Jazeera English (AJE) was born in 2006, it was the first among its sibling networks to bring a voice to the voiceless through its unique focus on the developing world in good times and bad.[1] But AJE was an extension of the trend the network had spearheaded from its first day of broadcasting as Al Jazeera's Arabic channel in 1996.

In Search of Independent Arab Journalists: Building Confidence, Building Journalists

Since game-changing media is not really about the state-of-the-art equipment or how sharply the cameras can focus but rather how independent journalists can be, Al Jazeera had a challenging mission from the start. In an environment as hostile to free and independent journalism as the Arab world had been for so long, independent journalists were a rare commodity. Salah Negm came in to direct the network news programming. A veteran TV journalist who started his career as a presenter for Egyptian radio in the early 1980s, he then moved to Radio Netherlands and later to the BBC's Arabic service. He moved on to Doha in 1996 to be one of the founders of Al Jazeera television and its first legendary director of news. Negm, currently the director of news for Al Jazeera English, recalls how hard it was in those early days.

"Recruiting, training and leading a staff of Arab journalists who came from countries like Iraq, Sudan and Egypt where they worked mainly for State-controlled newspapers, TV and media outlets was the toughest challenge I had in the beginning. The difficulty of the task was due to trying to mold those journalists' thinking into the freedom of expression and to tell the facts as they were without government intervention or self-censorship. The old habit of wait-to-follow-orders-coming-from-above was hard to overcome," he told me. "Coming from countries where authoritarian regimes indoctrinated them, it was a tough task to convince journalists that they have a paper in which they are free to write as long as they are factual, balanced and fair. The task was also about empowering journalists with knowledge so that they would be able to form their views and concepts freely. It was so tough it took us three to four years to change the collective state of mind of our staff journalists toward that direction," he explained. "Coming from the British Broadcasting Corporation [BBC] where I worked for many years as the executive news producer, I had the chance, along with a host of BBC journalists, to go from the BBC to Al Jazeera." Negm told me that this core group of journalists helped in recruiting and training the new staff of journalists.

It seems that the crisis runs deeper and is more profound than we might think. Many of the journalists who train and work in the west for free and independent media fall back to the mentality of controlled media once they are back in the Arab world. The interesting, yet sad, part of it is the geographical impact on how journalists perceive freedom of expression. They hold a sense of being free when they live in the west, but that leaves them when they go back to their Arab countries. I asked Negm about this strange mental transition, and he answered by saying, "Let me be frank, even that group of the BBC staff had a mixture of cultures; some of them moved to London fleeing their tyrannical regimes in the Arab world. Then, the BBC service recruited some from their countries, and others migrated from their homes in search of a better life or just more freedom. Each one of them had a different background. When they went back to the Arab world (Qatar) to work for Al Jazeera, each had their perception of how the new TV should be. Some of them felt the draw back to the same State-controlled media culture, and that was another challenge. Yes, the geographical impact you referred to is real," Negm confirmed.

"However, there is another aspect to that backward movement. Those Arab journalists who came back from the west to work for Al Jazeera had to navigate uncharted waters in the Middle East for the first time. Nevertheless, the determination of both Qatar's leadership and the top management of the channel enabled those journalists to sail that uncharted water and help them adopt a sense of freedom in their own part of the world," Negm pointed out.

As if the geographical impact was not enough challenge to Al Jazeera, they were later confronted by another challenge, which occurred during the first years of the network's existence. The other challenge was the impact of the exploding growth of military, ideological and political conflicts that have plagued those journalists' Arab countries in the newsroom and on the television screen.

"Another group of Arab journalists who fled their countries to the west because of their political ideologies sought to implement those ideologies once they were here at Al Jazeera. The way we dealt with them was by setting vision, values and principles. Therefore, if we detected an 'opinion-smelling script,' it was immediately discarded. Also, in the case of a one-credible-sourced fact-based script, we insisted that it must be balanced with another point of view. We have always had an efficient structure," the veteran journalist said as he charted out the network's methodology.

"Another aspect of the challenge was having a working group of only about 20 to 25 former BBC staff, which was not enough to staff and train a large organization like Al Jazeera. That group had to work around the clock on a daily basis for the first four years to prepare the environment for a functioning, independent free press. Eventually, the group who stood up for press values prevailed," Negm stated.

About the difference between managing the diverse newsroom staffs of Al Jazeera Arabic (AJA) and Al Jazeera English, Negm said the difference was mostly between the time he started with AJA in 1996 and then the later years when he led the AJE newsroom. "Human nature is the same regardless of nationality, but once you have good journalists, they will fly with freedom, objectivity, accuracy, balance and fairness—all the things good journalists would want, provided they can be certain they would not risk going to prison because of doing journalism the best they can. Therefore, it was difficult at the beginning, but once the right people were placed and principles embedded in the culture of the channel, there was no significant difference in managing the two different newsrooms—it was a straightforward operation," Negm explained. "However, proper staffing and good training do not rule out individual attempts to color news according to their ideology or for the sake of cultural benefit. Nevertheless, with a robust structure of check and balance and open discussion, these kinds of practices almost vanish. When you reach this point, there is no difference between running a newsroom in Arabic or English," Negm confirmed.

Ayman Gaballah is a seasoned journalist who started his career in Egyptian TV before he moved to work for Al Jazeera Arabic on the first week of its launch in November 1996. Gaballah eventually became a deputy chief editor and now is director of Al Jazeera Mubashir (AJM), which is a channel similar to the US-based C-SPAN TV but with more live coverage from all over the world, with a focus on the Middle East. As a deputy chief editor of AJA, Gaballah managed to lead a diverse staff, most of whom were coming from Arab state-controlled media. Additionally, they came from conflicting countries, yet AJA bet on revolutionizing stagnated Arab media and even employed journalists indoctrinated by dictatorships, who were brought up in ultranationalist journalism culture. Gaballah told me how he managed this situation. "We did not try to solve their countries' conflicts. Instead, we managed these conflicts all the time while in the newsroom by setting a golden rule: We all have to take off our ideological hats at the doorstep of the newsroom and keep only professionalism. Although it has been a big challenge, it worked. The mechanism we have been using to do so was the daily editorial meetings where we discussed, argued and explored all options before we set the perspective. As for dealing with journalists, many of them brought up on ultranationalist cultures with all the risks involved in rendering a polarized newsroom, we followed a strict code of professionalism to avoid such polarization," he added. "However, there were times when we had to deal with a few severe cases of polarization."

Establishing a New Culture of Diversity and Freedom of Speech

Ayman Gaballah went on to explain the benefits of a newsroom characterized by diversity and an open dialogue.

"For instance, during the coverage when fighting erupted in September 2007 between the Lebanese army and the Palestinian fighters in the River Bard Refugee Camp, some of our Lebanese colleagues whom we knew as open-minded and liberal, suddenly turned very nationalistic. Nevertheless, we made use of their view to try to be more balanced. In the meantime, it is noteworthy that human stories like civilians being killed ends any polarization," the seasoned journalist pointed out.

"Regardless of how anybody puts it, AJA since day one represented an unprecedented window of relieving distressed peoples in the Middle East who have been deprived of their basic God-given right of expression. Moreover, AJA was a game changer in the Middle East in the sense of being more progressive than all the politics in the region," Gaballah said.

"In the meantime, having such diversity of newsroom staff that represented all Arab countries enabled us to bring to light the peripherals of the Arab world such as Somalia, Mauritania and other countries that were long forgotten in the Arab and international media except for limited news in the time of famines and during conflicts. Moreover, and while being a game changer in the region, AJA became a serious competitor to international media," the man said.

Apart from relieving 350 million Arabs who have been deprived of freedom of expression for decades, AJA forced Arab dictatorial regimes after a few years of its launch to loosen their grip on their media. "Three years after contributing to launching AJA, I was part of launching two major media outlets: Abu Dhabi Channel, owned by Abu Dhabi Principality, and the Saudi owned Al-Arabiya. The two channels, and later others, came as a response to Al Jazeera. Although they came into existence as a result of AJA's challenge, they failed to match Al Jazeera in the spaces of the freedom it created. However, they had to expand a bit on the very narrow margin that existed before in their countries. AJA has been a revolution in freedom of expression in the region," Gaballah confirmed.

Emad Musa is a George Mason University graduate and now head of Al Jazeera English Online, who talked to me about how he saw the Arab world and later the whole South pre- and post-Al Jazeera, both politically and through the eyes of the media. "I was a young journalist when Al Jazeera started with two hours a day for two years before it became a 24/7 operation. I realized right from the start there was something different about that channel. Even during the first two years of the two-hour operation, it was evident that AJA was trying to find a place on its feet by experimenting with things that never happened in the Arab world before," Musa said. "As an Arab American born in the US to Palestinian parents who grew up on American western media, which became my field of study and I later worked for two US news agencies, this symbolized for me something almost unheard of. Never in my wildest dreams did I imagine that there could be something like this in the Arabic language, for Arabs, and coming from the Arab world, which has been consumed by dictatorships for decades," Musa added.

"Since I spent the first half of childhood in the US, and the second half in Palestine, where we could see Jordanian, Egyptian and Syrian TVs, I could see the massive gap between the media performance in the two worlds. As children, we used to make fun of what the state considers news and the way they presented it. Before Al Jazeera, it was not strange to watch an Arab head of state or king on the local news broadcast receiving the president of Zimbabwe, for instance, for 45 minutes on the news while classical music played in the background." Musa continued to draw a very dark picture of the Arab world before AJA and much less dark after it began broadcasting.

"Most of the Arabs who were born in the Arab world felt something was missing, but they had no alternative, especially before the technology of satellite dishes was introduced. For my part, I felt something was inherently wrong. Even in the first two years of its life when Al Jazeera aired on a two-hour schedule, they were different. They were agnostic to traditional things on local broadcasts, like breaks for the five calls of prayers and time zones," Musa said.

Al Jazeera Coverage of the Second Palestinian Intifada

Al Jazeera began to be a recognized household name in 2000 with its unprecedented coverage of the second Palestinian Intifada, during which it provided the first Arabic nonstop live coverage for hours, which kept Arab viewers captivated. That coverage was the dividing line between what had come before and what TV was now showing Arab viewers. For the first time in their history, what was happening in their backyard was broadcast live and uncut. However, that was not all.

"Al Jazeera live coverage of the second Intifada in 2000 provided the whole world with something they had never seen in the Arab context previously. At that time, when Al Jazeera opened the call-in feature, the central phone exchange in Qatar crashed because of the heavy traffic of millions of Arabs who wanted to talk freely for the first time in their modern history. Arabs desperately wanted to breathe freely; they wanted to exchange views; in brief, they wanted to be heard. Finally, they found a channel that gave 350 million Arabs a chance to talk and be heard," Musa reminisced.

About other impacts of AJA, apart from providing a platform for simple people to voice whatever they were thinking or feeling, Musa says that, equivalant to the internet's information revolution in the west, Al Jazeera was the beginning of the Arab information revolution, long before the entry of the internet. "Also, Al Jazeera provided a huge, unprecedented chance for Arab journalists at the time to work as real journalists not just as pseudo-clerks for governments or intelligence apparatuses," Musa added.

However, many Arab leaders have failed to appreciate the priorities of free and independent media and gone so far as to ban Al Jazeera from their countries. An example of Arab leaders' censorship is what the Algerian President Abdel Aziz Bouteflika did when Al Jazeera had to cut from a prerecorded interview with him in 2000 to go live with breaking news. To date, AJA is still banned in Algeria. "Despite being banned in many Arab countries, AJA, with its huge viewership, has reversed that equation in the Middle East, where media run after government officials. Now government officials run after Al Jazeera," the veteran Palestinian US-born journalist says.

PRE-POST AL JAZEERA REALITIES

Before the military coup on June 3, 2013, Egypt took one of the most enlightened views with regard to "freedom of expression" and was a country cetainly more willing to allow expression than just about all other Arab countries. But that is not to say the country was a shining beacon of light. If we know how badly freedom of expression was oppressed at that time in Egypt, we will know how horribly it was oppressed in the rest of the Arab world.

An Egyptian International Election Observer Who Never Voted!

Five months before Al Jazeera was born in June of 1996, I was called to the Egyptian Foreign Affairs ministry and asked, or cajoled, to serve as an international observer for the first multiethnic/multiparty election in the former Yugoslav Republic of Bosnia. I understood the reason I was being recruited. I was familiar with the geographical and political map of the country because I had covered the brutal war there for *Al-Ahram* for four long years. Even though the Organization for Security and Co-operation in Europe (OSCE) had requested my service, I was still shocked to think of taking such a job. I, the journalist who preached and wrote about the necessity of democracy, along with millions of my fellow Egyptians, had never participated in any election because we knew they were all rigged. But I agreed to be an international observer for this important election. After two months of observing pre-election campaigns and the voting process, I was praised by OSCE for my performance.

While I still felt somewhat ignorant of the process, it made me aspire to the goals of democracy even more. I hungered for a democratic culture in my own country, Egypt, and my region of the Middle East.

A Confession on the Fourth of July 2010

Fast forward 14 years from the scene at the Foreign Ministry to July 4, 2010. After 30 years in power, it was time to pass the presidency from Mubarak the father to Mubarak the son. As such a move required an absolute majority in parliament, I was becoming more and more convinced that a popular "volcano" was about to erupt. My feelings of an imminent and significant change had reached their peak, and it was, strangely enough, triggered on US soil in the heart of Cairo. On that night of the Fourth of July, when Americans celebrate their independence, I was standing in the large courtyard inside the US Embassy in

Cairo, watching hundreds of Egyptian and foreign guests celebrate. Music by a marine band was playing over the speakers. I was on the outer edges of the celebration chatting with a long-time colleague. As we considered the opposition's strong performance in the 2005 parliamentary elections—they won 88 of the 444 house seats—I argued that the regime would yield even more seats in 2010 if they were sincere about a gradual transformation to democracy. However, my colleague argued that the 2005 parliament was an exception and stated that the regime was unlikely to repeat the same "mistake" they did in 2005 when they allowed the opposition to gain a majority in the first and second rounds. It was unusual for the regime to allow such an outcome, but for the third vote, they returned to their usual practice and rigged the final election results. My colleague did not believe that the regime would take that risk again.

At that point we were joined by a longstanding senior adviser of President Mubarak, a man who was being presented publicly as one of the most prominent Arab thinkers. After listening to our arguments, he stated in an arrogant and threatening tone, "Are you two fools? The regime will never risk losing elections; we rig them, and we will continue rigging them. We not only know how to do it, but, most importantly, we could be very nasty with whoever points a finger." After making such an outrageous statement, he disappeared into the crowd, leaving my colleague and me stunned. I said to my friend, "change is coming very quickly; I can see the revolution is moving in fast."

This shocking statement made me feel that our country was standing on the edge of a dark abyss. Listening to the arrogance of the man bragging about how the regime could commit a crime and get away with it, my uneasiness about the upcoming election grew. The 2010 election results were unprecedented for their corruption. Mubarak's regime provided the opposition with no breathing room, and the ruling party took all 444 parliamentary seats. Mubarak's regime had consolidated its power.[2]

When a regime gets to the point of bragging about its crimes, it means that the end is on the way. The burning unanswered questions I had were: What would precipitate the ending, and at what price? Also, who would pay the price? These questions haunted me, and I felt a huge change was about to hit Egypt in unpredictable ways.

Breaking this code of silence in the service of state power has been one of the main objectives of Al Jazeera, especially in its early years. The network provided the first platform for Arabs and Middle Easterners to exercise the great freedom of expression they had long been deprived of by dictatorial regimes. Providing the means and venue for open discourse and political debate has been the most extended humanitarian campaign in the history of the Middle East, and it has lasted for years.

Dictatorships, the Media, and Fomenting Conflict: The Loss of Independence in State-Controlled Media

According to our experience from the Middle East, the main reason for internal and international conflicts is not the media. Instead, it is the dictatorships that control both the public and private media outlets. In this dark environment, dictatorships use media as a tool to divide societies to achieve their goal of maintaining and consolidating power and wealth. So far, most media in the Middle East is blocked from maintaining a culture of independence, and journalists are prevented from challenging their official sources, never allowed to ask them tough questions. Without media independent from the state, there will be no chance of creating a decisive role for the media in conflict resolution, spreading tolerance, presenting peacebuilding as a legitimate theme, or narrating the idea of peaceful coexistence. In such an environment, there is no way to counter extremism.

One of the main reasons for the many maladies Arab countries suffer from, especially the multiracial, ethnic and sectarian ones like in Iraq, is the unfortunate way that each sector within the county thinks of itself as "The Nation" by excluding the others. The natural effect of using media to spread such a culture of exclusion, instead of a vision for unity, is what we have been seeing for decades, resulting in internal conflicts that claim the lives of millions of innocent people, conflicts that bring some states to the brink of collapse. Arab dictatorships, especially those either led or backed by the military, have used media strategies extensively as lethal weapons. As media play such destructive roles, they lose any sense of professional standards and certainly have no code of ethics as guideposts. On the contrary, most of the media under Arab dictatorships are like runaway trains, spreading hate speech, intolerance and extremism, and have abandoned professional values that propose such things as accuracy, fairness and balance. This is the fast track to political, economic and social catastrophe and chaos, which has led to tidal waves of refugees spilling across the Mediterranean.

THE CODE OF ETHICS IN JOURNALISM

To most people, when they hear about a code of ethics in journalism, they think that it is about restricting journalists and forcing them to be nice and well behaved. This is a naïve and very superficial idea about the ethics of journalism. Rehabilitating a culture of journalism that reinforces adherence to a strict code of ethics is actually about enhancing professional skills, which are part and parcel of the fundamental values of journalism. A robust code of ethics is synonymous with professional journalism. A real professional journalist cannot be but balanced, fair, accurate and factual. All these values are the antithesis of hate speech, intolerance and extremism.

Breaking the Code of Silence: The Peoples' Voice

Imad Musa, head of Al Jazeera English Online, spoke to the role the network has played in opening up a public dialogue for the first time in the Middle East. "Indeed, it has been the longest humanitarian action operation in history. What drove Al Jazeera from the lowest level to the highest level has been its struggle to let everyone breathe freedom as a linchpin to equality and ridding the region of tyranny. The way to achieve this goal, according to AJA's prescription, has been to allow people to access accurate information. As far as I know, no other media played the same role the way AJA did.

"Al Jazeera stood up for the poor and the voiceless and produced media coverage for millions who never [had] anything like it before, despite their cultural heritage and their sufferings. Finally, they found someone who said: Yes, you count, and you have things that count; you represent the majority. While doing so, Al Jazeera did not turn its back to the officials and the power centers, but it provided another platform for the masses of people, opening up a debate between the *human centers* and the *power centers*," Musa said.

"For the first time in modern Arab history, we started seeing civil debate instead of live ammunition shootings. By doing that, Al Jazeera created a sphere that never existed in the Arab world before—a place to question and express skepticism by understanding that not everything we were brought up to believe is real, and we must question everything.

"Al Jazeera's motto at the beginning, 'the opinion and the other opinion,' was tremendously influential. AJA was the spark, and as a young journalist, I wanted to be part of this. For the first time in my life, something came out of the Arab world that I have been proud of; the only brand made in the Arab world that was known in the east and the west.

"At the beginning, there were 300 employees with less than 100 journalists who had a fire in their bellies armed with different vision, open sky, and were faster than other western media organizations of which each one employed as many as twenty to twenty-five thousand staff members. We have been working on a shoestring budget in comparison with the long-standing 100-year-old western media organizations and still scooping them. Al Jazeera proved that despite the importance of big budgets and staff, you can always scoop your competitors if you have the right vision. When Al Jazeera was born with the vision and the mission of relieving the distressed, we worked out of a small one-story building that had no amenities at all; those came much later. When I first worked for Al Jazeera, I joined the Washington bureau that had only one head and an assistant. Later, I joined Al Jazeera English when it was launched in 2006. What amazed me during that time were the volumes of résumés of both Americans and American immigrants who said how proud they were to work for such entity coming from the Arab world. It was building the bases for a new society," Musa said.

"Nowadays we see more and more the return of the state media hegemony. The commercial media are part of the hegemony of the dollar. AJA never had that kind of pressure. The state media, on the other hand, are trying to make their owners look like angels. Al Jazeera for sure was built on a very different set of principles; we only want to release as much accurate information as possible. That is why I believe that viewers in the world in general and Arabs, in particular, owe so much to Al Jazeera, which never tried to cash in on what it was doing. The Global South owes much to Al Jazeera, which clearly showed that there is something seriously wrong in the world by the persistent repression of the fundamental human rights," Musa concluded.

Opening a Dialogue and Shattering Myths: An Accidental Meeting with Weisman

At the end of the 1980s and the beginning of the 1990s, I was assigned by *Al-Ahram* newspaper to cover Israeli affairs. As a young journalist, I made a considerable effort to try to produce original and sometimes exclusive stories. However, the task was complicated. My work was constrained by the norms and standardization of Egyptian news and the expectations of the journalism community. I was restricted from citing any Israeli sources directly. The reason for the norm has been to prevent normalization of relations with the Israeli government. But one day I was presented with an opportunity that allowed me to fulfill my role as a truly independent journalist. I jumped at the chance.

It was a winter day in 1989, and I was running down Talaat Harb Street in downtown Cairo, headed to Groppi Café, when I found myself face to face with Ezer Weisman. Weisman had served as commander of the Israeli Air Force during the Six Days War in 1967. Immediately, I was surrounded by both Egyptian and Israeli secret service agents, who had very hostile looks on their faces. But considering his notoriety as Israel's defense minister during the Camp David peace negotiations with Egypt in 1979, he was surprisingly cordial and accessible, and so I began a conversation with him. Somewhat surprised by his presence in this old neighborhood in Cairo, I asked why he was there. The story was simple, he told me. "I loved to come to this café and sit for hours when I lived in Egypt," he said. I asked when he had lived in Egypt.

"I lived in Egypt during the years 1941–1942 while serving as a fighter pilot with the British Royal Air Force (RAF) during the Second World War." He went on to tell me his best memories were of meeting his wife for the first time, "right here at the Groppi Café." She had served as a nurse in a British military hospital. After they met and fell in love, they were married and spent their honeymoon in Egypt. He continued, "Since we established relations with Egypt in 1979, every time I visit Cairo, I have spent time at the Groppi Café."

Though the interview was of a personal nature, it demonstrated the cultural connections, often overlooked, between two nations frequently at war. According to the parameters of peace journalism, finding mutual ties is one of the first steps toward a diplomatic dialogue.[3] But for me, doing my job the way it should be done brought severe problems for me within the Egyptian press community that maintained the boycott against Israeli voices. Throughout those years, it felt as if I were the only boxer in the ring fighting a team of heavyweights all alone. Then Al Jazeera stepped in the ring and joined me.

Musa, head of Al Jazeera English, detailed the network's policy of including multiple voices in the news frame. "Since its birth on November 1, 1996, Al Jazeera was breaking down many myths every day in the Arab world, including those about Israel, by bringing Israeli spokespersons and officials live on its screens for the first time in Arab media history. Al Jazeera said: listen, Israel is a fact regardless of what we think about it, a position that reflected one of the principles of Al Jazeera, which is engraved on the wall of its training center: 'The moment we respect others' intelligence, they will respect and endear us.' Later, when I joined Al Jazeera and talked to many of its bureau chiefs, I was told how ordinary people in countries like Afghanistan saved their lives by tipping them off about dangers that many governments did not share. The common villagers in a country like Afghanistan knew that Al Jazeera was doing the right work, good work, so they respected and embraced it. They provided it with scoops."

Indeed, the Saudis were the first to make a significant investment in media by establishing the MBC group in 1991. However, though the Saudi station broadcast mostly entertainment, it did air a one-hour newscast that was essentially an imitation of some of the US networks. But only in format, not content. There was no core value to it. This may be because, as one Arab media scholar put it, "In the Arab world, privately owned satellite television stations are still indirectly controlled by the ruling national elites, whether through family relationships such as the case of MBC, or with a manipulated board of shareholders."[4] That is why 350 million Arabs could see a pre-Al Jazeera Arab history and then the era that followed. Something had fundamentally changed in 1996.

We will now turn to an in-depth discussion of the next phase of Al Jazeera's development, the evolution of the network into a renowned international news organization. The human perspective Al Jazeera presented in its coverage of conflict—showing the human face of war—offered a view that influenced global coverage and featured the peoples and voices of the Global South.

Notes

1. Philip Seib, *Al Jazeera English: Global News in a Changing World*, New York: Palgrave Macmillan, 2012. edition.
2. For a thorough discussion of the political corruption in Egypt that led to the Arab Spring, see Sahar F. Aziz, "Military Electoral Authoritarianism in Egypt," *Election Law Journal*, Volume 16, Number 2, 2017 "parliamentary elections were merely political theater to appease the regime's Western benefactors." p. 285
3. Dialogue in Peacebuilding: Understanding Different Perspectives, Dag Hammarskjöld Foundation, Uppsala, Sweden. https://webcache.googleusercontent.com/search?q=cache:NfF0JSO6MxgJ:www.daghammarskjold.se/wp-content/uploads/2019/10/dd64-dialogue-web1.pdf+&cd=2&hl=en&ct=clnk&gl=fr&client=firefox-b-1-d
4. M. M. Kraidy, "Arab Satellite Television Between Regionalization and Globalization," *Global Media Journal*, Volume 1, Number 1, 2002. http://repository.upenn.edu/asc_papers/186

2 From South to North
Reversing the Flow of Information While Covering War and Disaster

Phase II: From Time to Covering Wars

In 1999, I was assigned to cover the brutal war in the Democratic Republic of Congo (DRC). Unlike what most people thought, that war was not an internal one as much as it was a regional war, where all Great Lake–area countries were involved directly by having combat troops inside the DRC. The DRC conflict appeared on Al Jazeera screens but mostly through the work of news agencies. It was not until the 9/11 attacks that Al Jazeera leaped into the international arena with unique coverage of the global war in Afghanistan. When the network focused on the human side of the war and the brutal effects of bombing in civilian areas, Al Jazeera's presence as an international television broadcaster took off.

Al Jazeera's coverage of the war in Afghanistan was unique for many reasons, one of which was the network's access to the theater of war. Al Jazeera had six crews and direct access to the most dangerous and vital positions on the front lines, which were off limits to other international networks. Also, during the coverage of the war, Al Jazeera had almost exclusive access to the most influential sources of the conflict, unlike all competing networks. As a consequence, Al Jazeera's footage and logo were featured on the broadcast screens of many other international networks. It became the primary source for exclusive battle news. But such a feat would not have been possible without the steady development of staffing and training of field teams over the course of several years leading up to the conflict.

After Afghanistan and the 2003 invasion in Iraq, the war against Lebanon in 2006 and later in Gaza, Al Jazeera's reputation grew, and its unique global perspective began to take hold.

The network's unique perspective developed over the years, and it fundamentally changed the way news coverage of conflict was produced. Al Jazeera shifted the core of production from newsroom-based reporting to field operations and on-the-ground documentation of conflicts. The channel's management named Wadah Khanfer, a field reporter who headed the Baghdad bureau in September 2003, the new managing director and later the director general of the network. At that time, Al Jazeera had grown to more than ten channels, and it was fully committed to this new style of journalism and a new phase in its development. Coming from a field operation background, Khanfer directed resources to the peripheries where many people and communities felt the pain of war and away from of regional power centers. In other words, the channel became aware of the necessity of being *human-centered* rather than *power centered*. From this human-centered perspective, the view of war looked very different. Being on the ground with the people led the network to adopt a humanitarian perspective to conflict.

I asked Salah Negm, who directed the newsrooms of both Al Jazeera Arabic and Al Jazeera English, about these early developments and if he had been mindful that reporting the human aspect of war would enhance the work of humanitarian actors. Was it the network's mission to help alleviate the suffering of the distressed? I asked. He answered this way:

"In the line of news coverage, I do not like to use terms such as humanitarian action or crises. However, when I started working for the network in 1996, there was a different political environment than it is today both regionally and internationally. At the beginning of my work, the biggest humanitarian crisis was the serious lack of freedom of expression, democracy and almost total absence of the value of human beings and their lives. Therefore, I consider the first phase of AJA, from the date of birth in November 1996 until the beginning of the new millennium, was to change that de facto reality by raising the logo *the opinion and the other opinion*. The way to it was through establishing a media that told facts and provided platforms to all opinions and parties to reflect balanced views. By doing that, AJA was affecting positively the *biggest humanitarian crisis* in the Arab region and the Middle East *that both were entering a new millennium without having touched the 20th century's values of democracy and freedom of expression*. I believe that AJA made a huge difference in that respect," Negm said.

"However, seeking success was another motive behind what we have been doing. In our part of the world, all TV channels before AJA, usually supported by governments, sought success through focusing on news of presidents, leaders and the centers of power, but none of them investigated the lives of ordinary people. Our idea of success was different; to be successful was to talk to the people. You cannot talk to people through official statements. You must go and talk to that working man who is pushing a cart on the street to make a living. That man represents millions like him in the Arab world. If this man sees his story on the screen, other millions will also see themselves. This approach proved to be successful," he explained.

Al Jazeera, the BBC and CNN

"Technically the BBC has been the most accurate and CNN has been the fastest. At AJA, I thought that one alone was not enough; you must be the most accurate and the first to break the news. Personally, in the news, I do not believe in the saying 'better late than sorry.' Instead, it should be: If you are late, you should be sorry. In the meantime, you cannot be the first and wrong because if you are the second, you are going to be the last. As for the content we believed to be successful, you must reach for the ordinary people and have the best sources among the power centers; the two must go together. That is what we worked for, and it was the right formula. So, call it humanitarian or whatever, *Al Jazeera has been a media industry aiming at empowering people with information*," the seasoned journalist asserted.

Changing the Way War Is Covered

From 2001 until 2010, Al Jazeera adopted the phrase: "We Arabs and Muslims versus the world." Later that slogan developed into "We Arabs, Muslims and global South versus the world." The main characteristic of the change Al Jazeera would bring to global news was providing a unique focus to coverage of the wars in the Middle East.

I asked Negm if the strategy to reach out to all parties in a conflict had an impact on humanitarian action or the alleviation of human suffering caused by war. Again, Negm had a surprising answer.

"I left Al Jazeera in August 2001 and went back to the BBC. Therefore, I am not familiar with the beginning of that phase that you just mentioned. However, I will tell you about my experience with the coverage of Operation Desert Fox that was launched by the US and the UK under the Clinton administration in 1998."

The western media coverage of Operation Desert Fox[1] mimicked the coverage of the First Persian Gulf War and the US bombing of Iraq in Operation Desert Storm in 1991. US TV coverage presented the air strikes as precision bombing and the weaponry as "smart." Following the press briefings of military leaders, news media claimed that the "surgical" strikes caused almost no civilian casualties.[2]

"During that time, the practice of press embedded with the military was becoming fashionable. Nevertheless, during that war, all western media were expelled from Iraq except for AJA. We covered that war from the field in Iraq; the central command in Florida; Washington, DC;

and from the headquarters in Doha," Negm pointed out. "Our comprehensive coverage did not just show bombs being dropped and missiles launched, but we were also there on the ground showing what happened in the aftermath of the bombing and how it destroyed lives; human suffering was featured in our coverage. Of course, the Americans did not like that. However, being there on the ground pushed the Americans to be more prudent regarding bombing precision and the number of [dead and wounded] from what they call collateral damage. It made them think more about the embattled innocent people because we made it possible as much as we could for the world to be on the ground with us through the coverage," Negm said.

"That was the first time in the era of TV such coverage was introduced. Can you imagine if there were cameras on the ground right after Hiroshima was nuclear bombed? What could have been the impact on the public opinion? Do you think the second bomb was going to be dropped on Nagasaki?" the veteran journalist asked rhetorically.

Bearing witness to war's casualties and documenting the results of bombing civilian areas were the network's most significant influences on the rest of global coverage of conflict.

I spoke with another news director on the history of Al Jazeera as it became the primary source of conflict news internationally. I asked Ayman Gaballah, director of the Al Jazeera Mubasher Channel, whether coverage helped in relieving the people suffering the agonies of wars. He both confirmed and complemented Negm's view.

"If AJA were not there, history would have been written, or to be more precise, miswritten in a completely different way, and that is undisputable fact. From day one in 1996, AJA did not only break the news, but also it has been writing and recording the actual history, mainly of the Middle East and to a lesser extent other parts of the world. I remember when the first spark of the war on Gaza in 2008 happened. It was a shell fired by an Israeli soldier that killed 23 Palestinians. That morning all TV news in the region had their morning shows on. Our main competitor, Saudi-owned Al-Arabiya channel, had a gentleman talking about women['s] makeup. They did not want to interrupt their morning show and break the news. Rather, they kept that make-up gentleman on the screen for more than an hour after the incident took place, whereas we at Al Jazeera went immediately live with ongoing coverage. CNN, for its part, had the flash on their news bar. Finally, after more than an hour, Al-Arabiya did not take down the make-up morning show; instead, they split the screen between the aggression on Gaza and the make-up. I am sure: if it were not for AJA coverage, they would have probably carried on with the make-up show, ignoring the spark of the war on Gaza," Gaballah said.

Expanding the News Frame and Documenting Historical Events

"AJA forced other media both in the Arab world and worldwide to cover events they used to drop before. Meanwhile, Al Jazeera has constantly covered just about all events especially in the Middle East aiming at not only breaking the news but also building a complete repertoire of history for the future generations. Imagine if we were not there, we would have had massacres happening without knowing about them for days, weeks or do not even hear about them at all," the director of AJM pointed out.

"Documenting history is part of Al Jazeera policy, vision and mission, which stipulate AJA as a channel from the Arab world with international perspective upholding diversity and promoting democracy and openness. That vision and the mission have been pioneering not only in the Arab world but in the whole of the South and even among the marginalized in the North," he added.

The Voice of the Global South

Talking about the moment when AJA moved from being a voice of the voiceless in the Middle East to being a voice for the Global South, Gaballah expanded the definition and meaning of the word "South." It is used as a metaphor for the disenfranchised in the world geographically located both in the south and the north. He also asserted that Al Jazeera's mission and vision have become global trends. Al Jazeera has become more prominent on the world stage as an actor and an entity that has become a symbol of freedom, because of the repeated physical attacks on AJA staff over the years. Its correspondents have been harassed and detained, and many have been killed. Al Jazeera's overseas offices have also been bombed.

"Since that time AJA started to be recognized as an icon, a process that many people who never worked for AJA significantly contributed to all over the world from South Africa to Latin America, Asia and Europe. They all defended AJA, making it a phenomenon. It exceeded the idea of being just a good TV channel and became a *symbol* for the people who aspired to freedom," Gaballah concluded.

Breaking Down Hegemony and Reversing Media Information Flow

In 2007, a Commonwealth Summit was underway in Uganda. That conference was going to be exceptional since the Queen of England was going to open the summit in the capital city of Kampala. The Queen's agenda included a visit to an HIV orphanage as part of her humanitarian relief efforts. All international networks focused on her visit and the subsequent summit. Al Jazeera made the editorial decision that the main focus of their coverage would be HIV in Africa, the continent's ongoing tragedy, and its struggle against both the epidemic and the profiteering of the pharmaceutical companies trying to cash in on the catastrophe. Her Majesty the Queen's visit therefore became an entrance story to the main reporting, which was AIDS in Africa. Al Jazeera's decision influenced many other major western news broadcasters, including the United States, and the network's vision moved international news reporting in that direction. The in-depth reporting of a human-centered orientation in the news reporting was catching on.

Before the birth of Al Jazeera, we could see the enormous influence of western media on local Arab media to the extent that media scholars and news critics described the impact as western media imperialism. In the meantime, the Global South had long been virtually ignored by western media, with the exception of the standard stereotypic media tropes that often present the less wealthy countries of the world as synonymous with famines, droughts and epidemics.[3]

The questions I wanted answers to revolved around Al Jazeera's true influence on changing that equation. Did the network have such an impact on international news that it led to the rethinking of the media imperialism[4] perspective? If so, was it the result of a mindful strategy?

"Let's not use that expression, western media imperialism. Back in the seventies a famous report was issued titled the McBride Report.[5] The report revealed that the western news agencies, TV networks and all other media maintained control of the flow of information from the west to the south or east, even if the news was about the latter. For example, if you were running a local radio station in the east or south, you must get the news of your hemisphere from Reuters in London," Negm said. "Our objective in AJA right from the first day has been to reverse that flow of information. I do not bother myself with such expressions like 'media imperialism.' Instead, I think about big corporations in the west monopolizing the news when other people must be heard, and other perspectives of the news must be shown. We had to be strong and resourceful to be able to reverse that situation, and to be a source of news to western media, and that objective was achieved," Negm confirmed.

"In the beginning, we needed two years to strengthen our presence on the ground in many regions and countries in the south, namely the Middle East and Africa, where we had to reverse the flow of information, essentially becoming just about the only source for news from the south with a southern perspective. The flow of information must be a two-way-street. To do so, we had to spread our news-gathering operations, including offices and correspondents, into two distinct regions. One of the things that helped us to do that successfully was the fact that we as a network belong to the South. Having been treated as such, we were received differently by both the peoples and leaders of those countries. Our correspondents went to countries like Zimbabwe without carrying the same baggage carried by correspondents of CNN, NBC, BBC and other western media that brought the legacies of imperialism with them. Our correspondents went there with no preconceived stereotypic views, allowing the others [people and counties] of the south to reflect their views freely, but we did not stop challenging them. Therefore, they favored us," Negm revealed.

The 2002 War in Afghanistan

According to its former director general Wadah Khanfar, Al Jazeera wanted to invest in those countries rather than simply deploying temporary teams like many other western organizations had been doing. That is why and how AJA succeeded in providing such unique field coverage of many wars, including the invasion of Afghanistan, Iraq, Gaza, Lebanon and many others. In Afghanistan during the invasion in 2002, AJA was the only one there. "That is how we succeeded in breaking down the western media information hegemony," Khanfar said.

"Breaking the hegemony of western media was one of the greatest successes of Al Jazeera, and for the first time, mass media information was reversed. In the first ten years of its life, AJA was watched in cafés, homes, hospitals, bus stations in many countries such as Pakistan, Malaysia, Afghanistan and many parts of Africa. The amazing phenomenon that was AJA was watched all the time in these continents although their populations are non-Arab speaking," Emad Musa, head of Al Jazeera English Online, told me. "However, peoples in these countries have had held two sentiments at the time. The first is: why should we turn to the television of former colonizers? The second feeling was: although we do not understand Arabic on Al Jazeera, we should support what they are saying. Another reason was that networks like CNN, ABC, CBS, and others would show you where missiles took off when Al Jazeera showed you where they landed; the difference is huge," Musa continued.

What Musa said confirms Salah Negm, AJE's director of the news. He explained how and why Al Jazeera was given access to a permanent presence in countries where many western news organizations were denied.

"Although the Americans were constantly cursing AJA, their military used its information all the time. Regardless of being kosher or not, Usama Bin Laden was a newsmaker. His interviews with Al Jazeera were bought from the channel by many other international news distributors, including the Americans. The future generations will forever have AJA to thank for recording such detailed history from all aspects so as to present a more clarified and detailed version of the world order and events of the time. Without Al Jazeera, all that history would have been in the hands of self-appointed gatekeepers. Instead, AJA streamed it live unfiltered all over the world. Al Jazeera exercised real flow of information," Musa confidently said.

The Marketplace of Ideas

There is no doubt that freedom of the press was a result of a free entrepreneurial democracy where the free flow of unfiltered information was one of the essential sinews of these societies. However, and for the last 20 years, Al Jazeera has become the best representative of this open marketplace of ideas. Today, other media, many in the west, insist on practicing the gate-keeper kind of journalism, a journalism infused with nationalism and the priorities of the security state.[6]

"AJA was on the ground during the Iraq war showing Americans' unedited encounters of war, in which US soldiers were being roughed up, while in the US they were showing the sterilized version of the war. AJA has been showing the ugly face of wars, not just a video-game kind of war. Al Jazeera showed that there is no such thing as a hair-precision surgical strikes and smart weapons. Instead, it showed that wars are only about killing. It showed that there is no such thing as 'collateral damage.' Instead, innocent civilians are being eviscerated; there is no such thing as nice weapons. Instead, there are only lethal ones. That was one of the significant achievements of Al Jazeera, and it is not by any means an insignificant achievement," Musa stated.

"A lot of Al Jazeera's content was translated into English and shown on western media. It helped considerably to establish the anti-war movement in the west and especially in the US. The American movement's leaders had exercised their freedom of expression on AJA, while they could hardly do it in the land of the free. One example out of many was when AJA put on live the massive million US protestors gathered from all over the country in the Mall in Washington, DC, demanding to stop the war in Iraq. The reaction from millions of our Arab viewers was amazing, and it was all about how great to find out that the American people were not a nation of cowboys mongering for war. It was Al Jazeera that showed the Arabs that American people are not monolithic and that most of them are peace loving. What Al Jazeera did at the time was amazing," the Arabic US-born journalist said.

However, Al Jazeera had not only stood up to tyrannies and showed the ugly face of unjust wars, but it also had drawn the attention of the international community to humanitarian emergencies. Such attention drawn contributed to mobilizing the world public opinion, and therefore world leaders, into significant humanitarian action in order to alleviate the brunt of suffering of millions of human beings. Al Jazeera did that by being there before any other media to cover the tsunami in Indonesia, Thailand and other countries and, more importantly, by staying after the first shock when everyone else left. AJA did that again when it saw a disaster in the making in Somalia and East Africa nine months before the famine-drought struck and opened a permanent bureau there to hit the ground running once the disaster happened.

Of course, there are other examples of Al Jazeera being ahead of its competitors to draw the attention of the world to many crises that either happened or were in the making. Such a pivotal role played by Al Jazeera has been recognized several times by major international relief agencies, including those within the UN.

"Free, independent media, in most cases, cannot prevent wars but definitely can unmask their ugly face in the hopes of putting pressure on politicians who make these decisions to go to war. As for the role that Al Jazeera has been playing in bringing the world's attention to natural disasters, it is the highest hope of good professional journalists that they could improve the lives of others. However, I think that hope lies in the subconscious of the journalists while what lies in the conscious is to get as much as possible of the most accurate information as quickly as they can. Yes, we have to give credit to Al Jazeera for investing so much in bringing attention to long-neglected and ignored regions and countries and their peoples' sufferings because of natural disasters," Musa stated.

Institutional Agility and Covering Humanitarian Disasters

Apart from permanent bureaus providing coverage, in many cases, special assignments must be commissioned to areas where there are no offices. The decision process of such assignments in most networks is relatively lengthy and goes up the ladder to higher management. The process within Al Jazeera, especially when there are humanitarian emergencies, is short, quick and effective. Both the planning and assignment sections are given a free hand in dispatching correspondents and crews to any humanitarian crises, either human-made or natural, without the need for higher management approval. By doing so, the network has not missed one single coverage of any humanitarian crises from Haiti through sub-Saharan Africa, Latin America to Asia, up to the North Pole during that row over the ice-melting phenomenon and down to Antarctica (South Pole), wrapping the whole globe with its crews on the ground par excellence.

"The mission has always been to connect the Arabs, who have been left out by media for so long, with the rest of the world, and later when Al Jazeera English was launched in 2006, it was to even reverse that flow," Wadah Khanfar, former director general of the network, said.

However, to wrap the globe, an essential question must be asked: did Al Jazeera, in comparison to its western media competitors, need more of a newsroom-based operation or a field one?

Salah Negm, said to be one of the best news directors in the world of TV, says: "The newsroom is the brain, whereas the field operation is the muscle; to have a complete being, you must have the two. In the newsroom, you think about news globally, whereas the correspondents usually have a more focused look in the country or/and the region they cover. Therefore, one of the main tasks of the newsroom is to prioritize news, coupled with very strong news gathering regarding agile correspondents and crews.

"Despite the long history of networks like BBC and CNN and the fact that they have more overseas bureaus in comparison to Al Jazeera, they have a structure that slows their movements most of the time. Deployment teams of these networks are bigger, and the decision process takes longer that must go up through eight different levels to reach somewhere. As for Al Jazeera teams, they are smaller, more agile, more flexible; the decision is just one word, and the diverse nationalities of the staff enable us to choose the right people to be sent to the right places. Based on this diversity, along with other factors, Al Jazeera made a difference," Negm explained.

"Let me use the US–North Korea summit in Singapore held in June 2018 as an example. CNN sent a team of a 100, and the BBC sent a team of 120, whereas we at Al Jazeera sent a team of 20 to Singapore, the venue of the summit. When we look at the coverage, you would not be able to see the difference in the quality and the comprehensiveness at all. The difference is about the quality of the correspondents, flexibility and the ability to preview what is going to happen, be prepared and plan for it. In the meantime, the newsroom, which is the brain, decided to send teams to Iran, Israel and Korea, the three renegade nuclear powers, to give the coverage more context, and this is how you exceed the competitors," Negm asserted.

Qatar and Al Jazeera Under Siege

Since June 5, 2017, Qatar has been witnessing one of the most severe crises in its modern history when Saudi Arabia, United Arab Emirates, Bahrain and Egypt slammed a sea, land and air siege into the country, resulting in unprecedented distress to its own citizens and other internationals living there. The besieging countries had a long list of demands. At the top of the list that detailed 13 demands was shutting down the Al Jazeera network.

Never has a country been put under a siege because of a TV network, let alone one that has received worldwide praise, acclamation and international awards. To make such a demand demonstrates the power of the freedom of speech, and Al Jazeera represents an information revolution in a region whose dictatorships believe that information will empower the public, so much so that it will lead to them to demand democracy. The question is how has Al Jazeera, which played an essential role in enhancing the public awareness of humanitarian crises, approached this one with a price on its head?

Ayman Gaballah, director of Al Jazeera Mubashir, answered on behalf of the channel he directs, saying that they have tried to be as objective and balanced as they can in covering that event.

"The very nature of Mubashir as a reality TV helped us to be very objective and balanced. We provided a platform to Qatar as well as all besieging countries, sometimes even more balanced than we were with other events," he said.

Bearing in mind that the siege has been the most severe threat Qatar has faced in its modern history, I had to ask Director Gaballah if he has been given directions by the state or any of its departments on how to handle the crisis.

"One of the main aspects of the strength of Al Jazeera has always been the ability to show what the others cannot. We have kept this point of strength throughout the crisis. Don't forget that the list of demands of the besieging countries included shutting down the network. The best defense Al Jazeera could do was to show others' criticism to it. We, the directors, gave instructions to producers never to cut off any call-ins in which some callers insult Al Jazeera or Qatar. On the contrary, we encourage the producers to be very tolerant, rational and patient with these callers, just as they would with callers who praise us," Gaballah proudly said.

Knowing that Al Jazeera has stood up for the poor and the distressed throughout its 20 years, it will be interesting to know how it dealt with the disinformation machine of the besieging countries during one of both Qatar and Al Jazeera's worst crises in their history from the perspective of a US-born Arab journalist.

"The best thing they did at Al Jazeera was to focus on the human cost of the illegal siege. Also, the best thing was to use the right term 'siege' and not what the besieging countries used, as a 'blockade.' Moreover, Al Jazeera succeeded in cutting through all the propaganda of the other side and focused on the central issue: What is your problem? How could you rip apart and separate thousands of families of intermarriages? Although it has been in the eye of the storm by being on top of the demands list of the besieging countries to shut down, with its credibility and track record internationally and being less emotional, Al Jazeera did well for Qatar's cause legally, politically and humanely," Imad Musa, head of Al Jazeera English Online, said.

"On the contrary, despite being on the hit list of the besieging countries, Al Jazeera never stopped trying to interview the four countries' officials, who maintained one answer: 'We will not honor Al Jazeera by talking to it.' Even the Israeli officials have been smarter in dealing with media. Although they do not like Al Jazeera, they engage with it most of the time," Musa added.

"All good media hope for is to be an honest messenger and be judged as such. However, we do not get that chance most of the time in the Arab world. In such a rough neighborhood, it has always been kill-the-messenger policy. Under the circumstances, Al Jazeera managed to put a human face to many conflicts, although it was never immune to them. The most recent human face was to the Qatar crisis. Also, it played a crucial role in breaking down the besieging countries' rhetoric," the US-born Arab journalist concluded.

We opened this book with a parable about the values of good journalism: balance, fairness, accuracy, objectivity and neutrality were the Ten Commandments given to Moses on Mount Sinai. Historically, the modern form of journalism was a western invention, and so have been its values. In a region like the Arab world, where good independent and free journalism is a rare commodity, how far can we relate Al Jazeera to these religiously professional values?

"Listen, I am a simple man who thinks basically and mainly about news. I do not think about objectivity and neutrality as much as I think about accuracy and balance, which both enable me to evaluate news. Objectivity does not mean to take the right side, but it stipulates reaching the right side through objective means. It does not mean that you as a journalist cannot judge, but it means how do you reach that judgment and conclusion. It is about the approach, not the message. You have to be objective in exploring facts and factors to reach that conclusion. Therefore, if the objectively explored facts lead you to say this is a crime, you have to say it is, and this is what we have been trying to do all the time in Al Jazeera," Negm pointed out.

"There is nothing colorless in this world; the issue is how to reach the right degree of the color. As for neutrality, we should not and cannot be neutral. When we defend the human right, by definition we are not neutral," the man concluded.

Some of Al Jazeera's reporters and cameramen were arrested and harassed, even killed, since the launch of Al Jazeera 20 years ago. As if the persecution that many Al Jazeera journalists have been subjected to was not enough, Saudi Arabia, United Arab Emirates, Bahrain and Egypt decided to strangle the host country, Qatar, with the illegal siege. The siege resulted in many horrific stories of human rights violations of Qatari citizens living in the besieging countries, including being arrested or deported away from their families, especially the ones who have mixed marriages. How did the Al Jazeera Center for Public Liberties address this issue, which resulted in a severe humanitarian crisis for both Qataris and expats living in the country?

"With the accumulation of knowledge. The mission of the center evolved into spreading the culture of human rights rather than just publishing the violations. Therefore, we produced 65 promos; each one deals with the fundamental rights of humans including education, food, water, shelter, freedom of expression, the right of representation, decent wages and so on. Each promo is followed by a program detailing how this right is being violated all over the world and the means to keep it. During the siege of Qatar, we focused on the precarious situation of the humans rather than of the governments such as separating families of intermarriages, and there are many of them. Also, the Saudi's official decision to deprive Qataris, and most of the expats living in Qatar, from making a pilgrimage to Mecca and Medina," Al-Haj explained.

As if the grave violations against Qatar, its citizens and expats were not enough, the besieging countries' regimes imprisoned and tortured many of their intellectuals who refused to adopt the same hate speech against Qatar.

"We exposed the hate speech used by the besieging countries against Qatar's citizens and expats both on the official level and in their media. In the meantime, we made sure to immunize Al Jazeera of this hate speech, so we do not reciprocate with the same speech. In that respect, we asked the specialized international organizations that monitor hate speech in the media to watch our performance. Meanwhile, the center heads the crises committee of Al Jazeera on which all the network platforms are represented," the former Guantanamo inmate/journalist concluded.

We will return to the topic of the state of siege in Qatar in later pages. The next chapter details the theoretical foundations and practical structures that have allowed Al Jazeera to take a primary place on the global stage as a leader in documenting and reporting the plight of people and communities caught in humanitarian emergencies of all descriptions.

Notes

1. Jon Schwarz, "A Short History of U.S. Bombing of Civilian Facilities," *The Intercept*, October 7, 2015. https://theintercept.com/2015/10/07/a-short-history-of-u-s-bombing-of-civilian-facilities/
2. Robin Andersen, *A Century of Media, a Century of War*, New York: Peter Lang, 2006. Chapter 12
3. Suzanne Franks, "Reporting Humanitarian Narratives," *Routledge Companion to Media and Humanitarian Action*, Robin Andersen and Purnaka L. de Silva eds. New York: Routledge Press, 2017. "The narrative of humanitarian suffering very often relies on familiar stereotypes and fails to convey the complex underlying politics. The reporting uses instead frames such as 'primitive tribal hatreds' or resorts to explanations based upon 'natural disaster' when there are in fact complex underlying social and political causes to many crises and complex emergencies. There is a marked difference from the way that crises are reported when they occur in Western countries, where there is a far greater attempt to give underlying explanations and nuanced interpretations that take account of political factors and present the suffering victims with far greater context."
4. Oliver Boyd-Barrett, *Media Imperialism*, Newbury Park, CA: Sage Publishing, 2014.
5. The McBride Report, *Communication and Society Today and Tomorrow, Many Voices One World, Towards a New More Just and More Efficient World Information and Communication Order*. Kogan Page, London: Uniput, New York: UNESCO, Paris: UNESCO, 1980.
6. Max Blumenthal, *The Management of Savagery*, New York: Verso Press, 2019. Chapter 9.

3 A New Kind of Humanitarian Journalism
Partnerships, Coalitions, Research and Investigations

Setting a Precedent

After the massive defeat Egypt suffered in June 1967 at the hands of the Israelis, the political leadership of Egypt realized that one of the main reasons for the defeat was the fact that the military was deeply involved in politics, supposedly the monopoly of civilian rule.

The late President Gamal Abdel Nasser took strict measures after that time, including firing many of the army generals and trying others. The new generals and commanders were entirely banned from even getting close to politics. Despite those measures and many others, Nasser had suffered a humiliating defeat. Almost the entire armed forces were destroyed, and Egypt lost the Sinai Peninsula, which represented one-third of the country, to the Israelis. Nasser thought about establishing an independent military think tank apart and away from the armed forces. The think tank's mission was to come up with unconventional solutions to Egypt's military situation, which was considered dire at that time. Nasser trusted no one with this mission except his long-time confidante Mohamed Hasnain Heikal, the stellar chief editor of *Al-Ahram* newspaper, and so Al-Ahram Center for Strategic Studies was born in 1968, a few months after the Six Days War.

Based on documents released after the October 6, 1973, war, the center played a crucial role in reversing the military situation in Egypt's favor. To give an idea of the importance of the center's role, it is enough to know that even the battle orders during the Yom Kippur war were written in the center. Although it always worked under a dictatorship, the incentive of the whole country to overcome the humiliating defeat of 1967 secured independence for the center. When that incentive was gone, the prestige of the center fell drastically, both regionally and internationally. Under a new dictatorship, there was nothing to save the country, or the center as a part of it.

However, the fact remains that the *Al-Ahram* newspaper/foundation was the world pioneer of having such a think tank that housed scores of the best experts and researchers of the country.

Al Jazeera's Center for Studies

Forty years later in 2006, the Al Jazeera network established Al Jazeera Center for Studies to conduct in-depth analyses of current affairs at both regional and global levels. Its research agenda focuses primarily on geopolitics and strategic developments in the Arab world and the surrounding regions. Based in the heart of the Middle East and operating from within the sociopolitical and cultural fabric of the Arab world, Al Jazeera Center for Studies seeks to contribute to knowledge and sharing to formulate a better understanding of the complexity of the region. With an extensive network of distinguished researchers and a broad array of experts from across the globe, the center aims at promoting dialogue and building bridges of mutual understanding and cooperation between cultures, civilizations and religions.

However, as a think-tank extension of the Al Jazeera Network, the original idea, before it developed to cover a broader spectrum of missions, was to conduct research and build relevant, insightful, in-depth knowledge for the entire organization by groups of experts and analysts who are not under pressure of news gathering and coverage. Their job has been to analyze news and contextualize information for more complexity before passing it on to newsrooms.

"The center is a think tank located in the most turbulent region in the world, the Middle East. There are two main roles the center plays: The first is an educating one to Al Jazeera journalists by providing them with material which would enable them not to stop when and where the news ends but also to go to the length of what is behind and beyond the news. The second role is to supply both the newsroom and the training center with the needed material in its field of work. For example, now as we are talking, AJA is producing a documentary about World War I; we as a center are providing them with fact-checked academic material to be at their disposal. The first role is about enriching the knowledge of journalists and expanding their strategic thinking, and the second is providing material whenever needed," Dr. Mohamed Al-Mokhtar, director of the center, said.

"In other cases, all the network's platforms come to us with a ready-to-go program to be academically judged by experts in the center as a quality control process in which we judge information and whether it fits the context and news frame. Other times, they come to us to help in setting a context to certain materials. The third role is to provide the newsrooms with our experts and researchers as guests speakers. The fourth role is to contribute to organizing the network's forums and conferences," Dr. Al-Mokhtar added.

For example, when Queen Elizabeth of England made a rare visit to open the Commonwealth summit in Uganda that I mentioned in the last chapter, the Al Jazeera newsroom had a debate on how best they would cover that historic trip to an African country. The Study Center's view was that the visit to open the summit should not be the primary goal of the coverage as much as using it as an entry to focus on the long-forgotten disasters, epidemics and famines hitting the African continent and on the historical and moral responsibilities of the former colonial powers on the current realities of Africa. Also, a decision was made to use the Queen's visit to expose western pharmaceutical companies' greed when they take advantage of Africa's suffering to increase their profits by trillions regardless of the continent's soaring mortality rates. The center's view was to bring Africa and its chronic diseases to world attention, something that put international media in an awkward position. That was just one example of many of the contributions of the center.

The question is: How often does it happen?

"At the center, we do this job from time to time, whenever needed, in collaboration with the newsroom but not in the form of a decision taken by the center but rather as a result of brainstorming with the newsroom. The center is contributing to the production of two of the main shows on AJA: 'Lanes' and 'More of the Story,' and a representative of the center is always there during their editorial meetings. The reason for such greater involvement is that the nature of the two shows is more analytical," Dr. Al-Mokhtar revealed.

One of the fundamental roles of the Al Jazeera network from day one has been to provide platforms for millions of the once-voiceless in the Middle East and later in the Global South in pursuit of fundamental human rights. What role has the Study Center played, if any, in defending the network's host country, Qatar, during the siege brought by Saudi Arabia, United Arab Emirates, Bahrain and Egypt, especially because shutting down Al Jazeera was at the top of the list?

"The center has been playing a vital role in that respect while doing its best not to overact and strongly advising the network's platforms to be rational in reacting. Our primary focus has been researching while directing researchers to do their job objectively and not to be preconceived result oriented. We published two books on this matter in which we dealt with the siege by detecting the reasons and dissecting the mechanisms of the siege in the first book. The focus of the second book addressed how small countries like Qatar can resist and survive sieges by large, wealthy countries," the director of the center pointed out.

The Need-for-Change Journalism

Over the many years that Al Jazeera has developed into an essential global broadcast network, it has transformed the coverage of war, brought human suffering of those experiencing disasters of all sorts to TV screens and shown the victims on the receiving end of the world's most destructive weapons. It's also been a beacon of light to those who yearn for freedom. That has not happened in a void, and there is much to be said about the long discussions and of the role media can play in working toward social goals and a more humane and just world. We will now address the theoretical and practical forces that have led to the network to achieve such an important global status.

The Watchdog Press

American reporters like to joke, "Don't trust anyone but your mother, and whatever she says, check it." During the last two decades, most US reporters' work has been different. Even sources who say things that cannot be verified, and that would immediately appear as false if they were verified, are quoted in daily reports. This is surely one reason that news media are regarded by many of the public as lacking credibility. Put simply, they allow sources to lie, especially high-ranking officials who are quoted without challenge.

In an article titled "An Historical Approach to Objectivity and Professionalism in American News Reporting," Dan Schiller[1] revealed the point at which US reporters stepped back from dependence to more challenging journalism. Schiller stated then that the " 'quasi-institutionalized dependence' of reporters on official sources was weakened in the United States during the Vietnam War when it became apparent that officials were misleading the public and using journalists to do it." However, in an age of corporate conglomeration and the profit-driven news media, most reporters have lost the habit of challenging officials. One reason is that they do not have time to gather information to support challenges; most news organizations are understaffed. It is easier and cheaper to accept official press releases. Indeed, many international news bureaus once staffed by US media organizations were closed as news became increasingly monetized during the latter half of the twentieth century.[2] A second reason is that reporters in many places see themselves as guardians, not critics, of the state and its values.

In *Confessions of an Investigative Reporter*, Christopher George warns that, "Most reporters who claim to practice investigation are merely channels for sources who provide them with scoops. They find nothing for themselves; their stories are rewrites of reports compiled by government or civil society organizations."[3] In contrast, authentic investigative reporters do not accept something as true simply because someone, in particular an official, gave them a report. Instead, they verify the information through other sources and challenge information that cannot be verified. Media outlets frequently accept official reports that come from government sources without verifying their validity. This is especially true with conflict reporting and foreign policy issues. A well-documented case in the literature is the infamous white paper released by the Reagan administration that claimed that Soviet weapons were being supplied and transported from the Nicaraguan government to the rebels in El Salvador. Media reported these claims as facts, though they were not verified in the report.[4] The importance of a source to an investigative journalist is based on the validity of the information that the source can provide.

Most Arab media have for decades been dependent on unchallenged sources but also indirectly, and sometimes directly, employed by them until Al Jazeera was born in 1996. AJA introduced challenging journalism to the Middle East, journalism that arrives at a judgment that something should never have happened, that it must never happen again or, conversely, that something worthwhile has been unjustly terminated. In an era when the world is in great need of social justice, peace and security, this type of media has been called "the need-for-change journalism."

Media and Journalism as a Sociopolitical Force

The contemporary concept for "story actors" is *stakeholders*, defined as people who have an interest in the outcomes of a given organization, issue or situation. Mark Lee Hunter, "Story Based Inquiry: A manual for Investigative Journalists," *UNESCO*, 2011, https://unesdoc.unesco.org/ark:/48223/pf0000193094/PDF/193078eng.pdf.multi.nameddest=193094. If a journalist constructs a story without including or consulting stakeholders, they will rarely be able to comprehensively understand the meaning and significance of the subject of their reporting. One of the ways media and journalism augment knowledge in the service of social outcomes is to work with relevant stakeholders. This is the best way to help drive solutions to problems, injustices, intolerance and extremism. In short, stakeholders enter media not only as audiences. They also supply:

- Sources
- Logistical support such as help with travel, lodging, access and so on
- Protection via legal aid or political influence

40   *A New Kind of Humanitarian Journalism*

In the book *The Journalism of Outrage: Investigative Reporting and Agenda Building in America*, David L. Protess[5] and co-authors carefully examine how investigative journalism achieves results regarding prosecution, legislation and other action that leads to reform. In this seminal work, media and journalism together are seen as a social force. The authors identify two main models for achieving social goals that help drive reform. These models apply to stories from the departure point—research—right up to the final outcome—publishing. The models are as follows.

Mobilization: An Incomplete Model

Public outrage caused by the journalistic revelations of a scandal frequently results in actions by political and social leaders. When media expose corruption, political scandals and outrageous injustice, such revelations have the power to incite the public to anger, and that outrage can be heard. Of course, this is the legend of Watergate. Two courageous journalists exposed the corruption of a sitting president in the pages of the country's leading newspapers, and the president was forced to resign. It is noteworthy that the legend is not entirely accurate. Although public outrage at the unfolding scandal was great, reporting the illegal actions of President Richard Nixon would not have been enough without the support of a number of journalists, media outlets, judges, other politicians and congressional leaders critical of the president's actions and behaviors.

However, we cannot merely dismiss the mobilization model. Before Watergate, consumer activist Ralph Nader demonstrated the power of engaged, or advocacy, journalism to mobilize public opinion. Nader was not a journalist, yet he used investigative reporting in the long-form version of a book to change the face of the automobile industry in the United States. In his book *Unsafe at any Speed*, Nader[6] used the journalist method of mobilization to persuade the public that their safety mattered, and he had lasting effects.

The mobilization model began well before the emergence of muckraking journalism. In the late nineteenth century, campaigns by newspapers in the United States to achieve reform were called "crusades." Although perfectly aware this term has a repugnant connotation in the Arab world, because of its association with the brutal massacre perpetrated by the early crusaders in the eleventh century against them, we must clarify that in the Christian world, moral crusades have been practiced at home as well as abroad and are an accepted part of that culture.

Nellie Bly's *Ten Days in a Madhouse*[7] was the cornerstone of one such crusade—the determination to reform asylums. Her story captured critical elements of crusading journalism. For example, she merged facts with emotions, including shocking revelations and the graphic language of outrage, all leading to an explicit demand for reform. Some of these crusades, such as the decades-long campaign to end slavery, can be defended as necessary and just, despite the tragedy of the American Civil War that followed. Others, such as the crusade against alcohol that led to Prohibition in the 1920s, were counterproductive; Prohibition merely led to the emergence of wealthy criminal syndicates. Still others, including the crusade of the Hearst newspapers to launch a war against Spain in 1898 or the attempts by the administration of George W. Bush to justify an invasion of Iraq by the presence of nonexistent weapons of mass destruction, were cynical attempts to manipulate public sentiment.[8] Here we find the beginnings of what has grown into full-blown misinformation and been termed Fake News in the second decade of the twenty-first century.

We would argue that mobilization is at best an incomplete media and journalistic strategy if your goal is social justice and political reform. To achieve reforms, it is usually not enough for the media to simply outrage the public or to humiliate leaders through the revelation of unsolved problems or political corruption.

Coalition: The Ideal Model

Media can achieve results most effectively when they form coalitions with other social forces. It is noteworthy that isolated journalists rarely succeed in their pursuit of reform or justice. Other actors must be engaged, both to protect media outlets and individual journalists from reprisals and to succeed with progressive social change.

Some ways that collaboration takes place include sharing sources and contacts, research and data, and even access to witnesses. Though journalists must verify all the information they have gathered before relying on it, they can also obtain documentation, research findings and scientific data from a variety of experts and foundations. Whistleblowers and NGOs are among the most common sources of such material. However, journalists should not simply use a report by someone and rewrite it. Like any other source, partners in a coalition must be regarded as fallible and can act in their own interests instead of those of the broader public. Like any other document, authentication of third-party reports must be verified independently. Such forms of collaboration are becoming more common. For example, NGOs and other humanitarian organizations working in the fields of aid and development often compile essential data and share it with media outlets and journalists. Recently, Greenpeace adopted a collaborative strategy, even offering to pay print, TV and radio journalists to write reports. Investigative journalists working in different countries may also share information in this way to pursue international aspects of a story.[9]

In 2008, Al Jazeera began to build coalitions with humanitarian actors to augment its international coverage of conflict. I spoke with Wadah Khanfar, former director general of the network, about the growing need for crisis awareness and catastrophe prevention. He said, "Contrary to heavily commercialized media, we at Al Jazeera have always been aware of the importance of media's role in contributing to relief efforts for suffering people hit by either human-made or natural disasters."

He told me about the 2008 launch of Al Jazeera's Center for Human Rights, saying, "The center was preceded by a conference organized in Doha by the network in which most of the important human rights organizations and humanitarian relief agencies were in attendance. The conference was concluded with the signing of memoranda of understandings between the then newly established Center for Human Rights and these organizations about future cooperation," Khanfar added.

Since then, many international humanitarian organizations have praised Al Jazeera's news coverage and the role it plays in helping to ease human suffering by raising public awareness and mobilizing relief efforts during crises. Al Jazeera been decorated and awarded several times for such roles, especially when it comes to raising awareness about the tragic plight of refugees displaced in many parts of the world. In 2005, the United Nations High Commission for Refugees (UNHCR) awarded Al Jazeera for its coverage of the tragedy of refugees resulting from the wave of drought that hit West Africa during 2004–2005. Reporting began shortly before the drought hit, thanks to cooperation between the network and relief agencies. Such coverage played a crucial role in mobilizing international efforts to better deal with the tragedy. There are no other media networks that have such unique and effective partnerships and coalitions with humanitarian organizations. Following are some of the ways in which these reciprocal partnerships are structured and nurtured and what it takes to facilitate these important coalitions.

Sharing Powers and Rights

Different actors in a story may have powers that are not legally available to journalists. For example, the plaintiffs in civil or criminal procedures and their lawyers typically enjoy access to evidence that reporters cannot easily obtain. Moreover, the history of journalism contains numerous examples of reporters who collaborated with public officials eager to fight corruption. Pulitzer Prize–winning American investigative journalist Clark Mollenhoff began his career in the provincial town of Iowa City, where gangsters bribed politicians and got away with it. Mollenhoff formed a team with an honest prosecutor and a trustworthy policeman.[10] The prosecutor built the case, and the policeman arrested the malfeasants, and Mollenhoff provided leads and support through his published reports. The critical insight here is that journalists do not send criminals to jail. Police, prosecutors, judges and juries do that. Journalists are thus advised to identify honest people in each of these occupations and to consider how they can cooperate without compromising their ethics, reputations or objectives.

However, there is a risk in such cooperation. One risk is very well known. Journalists who seek to cooperate with public officials, businesspeople or social actors like NGOs may lose perspective and therefore their independence. Eventually, many such journalists openly share in the power that emanates from a public figure, as followers. In the end, the journalist may be blinded by that power and incapable of seeing clearly. It is essential for journalists to know their role. If they make alliances with other actors in the society, those alliances must be temporary, with a clear objective in mind. Once that objective is obtained, relationships with other actors must remain at arms length, which is another absolute rule of Al Jazeera.

Sharing the Story

A fundamental insight of the coalition model is that an isolated journalist cannot succeed on their own. It is easy to be ignored. Even if the revelations are powerful, they will soon be forgotten in most cases. Unless the media outlet has compiled sufficient information to support a series, the single story will quickly become old news. Equally important, a single media outlet or reporter can become vulnerable. Pressure will be put on editors, publishers and broadcasters to take action against a reporter who veers too far from standard reporting or provides an alternative to the dominant news frame. They may recant or apologize for a story or reassign the journalist. Such was the case with Gary Webb, whose three-part series that ran in the *San Jose Mercury News* about drug dealing and the Iran–Contra scandal was discredited, and the repercussions of his reporting eventually led to his death.[11] If an aggrieved party sues a reporter, judges and juries will note that no one seems willing or able to defend the reporter's revelations. Even if the case does not go to court, the reporter may come under attack by the targets of the exposé, their influential friends or associates and even biased media outlets and journalists.

A case in point is from my own experience of publishing a series of investigations from September to December of 1993, detailing the ways that corrupt ministers were able to seize a historical site on the Mediterranean to build a number of chalets. I shared the story with many stakeholders who had a deep interest in stopping this crime. The stakeholders included the governments of Italy and Greece, since that site belongs to the Greco-Roman era in Egypt. By sharing the story, I managed to build a coalition of corruption fighters that eventually forced the ministers to give up the land and drop the lawsuit they brought against me.

It is not enough to be right. A wise journalist will ensure that they are not alone in knowing the truth. To reduce personal risk, the journalist must ensure that:

- The facts of the story are not overstated.
- A viable path to the facts must be cleared for other journalists.
- Social actors can be engaged—as sources and as potential allies—before the story is broadcast or published by alerting them that a controversial story is coming.
- Equally important, nowadays, nearly all NGOs and other civil society actors have websites or other media such as publications or video streams. They can thus provide support and publicity for a story whether or not other mainstream media pick up the story. In some instances, these "stakeholder media" can have a more powerful influence than mainstream media.

THE BACKLASH TO FREEDOM OF SPEECH

Using these models not only maximizes the impact of the media but also minimizes the sociopolitical backlash against it, which comes in different forms, including:

Targeting Advertising Revenues

This tactic is designed to rob the media outlet of revenue. It will be employed as long as media depend upon advertisers. It can be highly effective, especially with media that lack a diverse and broad funding base.

For Al Jazeera, this tool for punishing independent media is obsolete, because from the beginning, it was never a commercial network and does not depend on advertising to generate its broadcast funding.

Audiences, Ratings and Commercialism

According to many Al Jazeera officials, the relationship between the network and its audiences is uniquely dialectical. Ratings serve as a detector to the public's pulse. However, in many cases, the audiences come *to* Al Jazeera because they trust its capacity to choose for them. It is a very long and sensitive process of building trust and credibility between the network and its audiences.

"For Al Jazeera, ratings are important but only to a certain extent, for it is also part of commercializing media, which in many cases infringes on audience' rights. The fact that Al Jazeera shies away from commercializing news makes sense. Imagine if ratings were not in favor of coverage of humanitarian emergencies; what would be the outcome in a profit-driven network? Would that negatively impact coverage? The realistic answer is yes," Wadah Khanfar said. "Nevertheless, in the case of Al Jazeera, and except for the news, most media aspects could be commercialized, such as sports and entertainment. These aspects of media work should sponsor news to keep it from being a mere reflection of the centers of power," Khanfar added.

The Threat of Lawsuits

Taking legal action against journalists and media outlets has become more common since the Watergate era in the United States and elsewhere. Whether or not a reporter has employed careful documentation and legitimate sources, the threat of such lawsuits can have a chilling effect because of the cost of legal action. The threat of legal action is used to dissuade journalists around the world. Though the US Supreme Court generally protects the rights of journalists to tell well-documented stories about the misdeeds of public figures, editors and publishers are increasingly worried about the possibility of libel suits.

Al Jazeera has increasingly become the target of libel suits over the last ten years. Most of the cases come from the Middle East, and their purpose is to intimidate journalists and the network from telling the truth. Independent fair and free journalism cannot thrive in a region dominated by dictatorial regimes, and Al Jazeera has become the enemy of those who seek to control opposition voices and public expression in general. However, in the case of Al Jazeera, this tactic has not stopped the network from engaging in hard-s, clear, well-documented investigative journalism, the type of reporting most likely to result in lawsuits. The network fights these censorious actions at a number of levels. It stipulates strict criteria for quality reporting that must be followed for even-handed professional journalism. In addition, if a case is brought, or a journalist is detained on politically motivated charges, Al Jazeera has a significant legal department to deal with such types of harassment for its journalists, which come mostly from Middle East dictatorial regimes.

Official Repression

In recent years, struggles between governments on one hand, and journalists on the other, have been escalating, leaving advocates of transparency and freedom of expression under attack. In Egypt, for example, the military obtained convictions for 43 NGO workers, including Egyptians, Jordanians and Palestinians, on charges of working for unlicensed institutions and receiving illegal funding in 2013. One of those NGOs was the International Center for Journalists (ICJ), an organization that provides training for journalists to enhance their skills. Because of my work with ICJ, I was one of those convicted. I left the country and have been living in exile since that time.

According to the Middle East Monitor, an organization defending press freedom in the Arab world, the number of journalists facing forced disappearance, arbitrary detention and corrupt trials is up to 100, and that number continues to grow.[12] The targeting and jailing of journalists around the world has dramatically escalated in the twenty-first century.[13] China has targeted anti-corruption activists. The list of journalists murdered in Russia has become considerably longer during Putin's domination of the country. Even in the United States, long considered a champion of freedom of expression, media has been targeted by the White House as "the enemy of people," and whistleblowers and journalists have been jailed. Journalists have also been attacked while covering the rallies of President Trump. These actions give license to authoritarian regimes worldwide to close down freedom of expression and freedom of the press.

However, this does not mean that reliable, independent journalism is coming to an end in the west, nor does it signal a loss of hope for the free press to be born in the Arab world.

CONTINUING THE TRADITIONS OF GOOD JOURNALISM

First, the adversarial tradition remains alive. Consider the many journalists who continue to risk their freedom, and their lives, in a region like the Middle East. On the institutional level, we must remember Al Jazeera and other publications such as *The Intercept*, an online news source created to expose the activities of the US National Security Agency, institutional examples of independent, robust journalism. *The Intercept* statement mission reads:

> Our long-term mission is to produce fearless, adversarial journalism across a wide range of issues. The editorial independence of our journalists will be guaranteed. They will be encouraged to pursue their passions, cultivate a unique voice, and publish stories without regard to whom they might anger or alienate. We believe the prime value of journalism is its power to impose transparency, and thus accountability, on the most powerful governmental and corporate bodies, and our journalists will be provided with the full resources and support required to do this.[14]

In other words, official repression has not reduced the desire for journalists to speak truth to power.

From Mobilization to Coalition

Investigations into humanitarian crises may involve either the mobilization model or the coalition model and, to maximize the impact of a given case, both models may be employed.

I asked Salah Negm, AJE director of news if, in the course of his work, he used any of these models.

"My basic concept in news coverage is that we must not adopt a cause. The main criteria for deciding whether or not we are going to cover an event is its importance. Some stories are good for only a day, while others unfold over time. Based on answers to questions such as: How many parties are involved in the story? Does it have repercussions? How much is it revealing? If all answers are positive, then we decide to own the story by *mobilizing* our resources both in the newsroom and in field operations to cover it from all angles, which means more space and time. By doing that, the story will be dominant." Negm went on to note that mobilization in this case is defined by the amount of resources assigned to the story. In this case, the network has decided to "own" the story.

I followed up with another question to Negm, asking if, when resources are assigned and a commitment is made for professional reasons, are you not sending a message inadvertently to viewers to mobilize as well?

"Yes, that might be the inadvertent message. However, there is an intended message, which is owning the story. Of course, sometimes it happens that some journalists in the newsroom argue that we should vary news to keep the audiences as broad as we can. To those journalists,

I always say that if we owned the story, some of the audience might switch to other channels, but they will always come back to us to follow the story we have committed to. There are many examples in which that part of the audience kept coming back to us, such as during the coverage of the tsunami in 2006 when we were the main source of the news not just to our viewers but to many international media as well," Negm said.

However, as frequently happens, when there is more than one critical event to cover, news editors must decide to give more time and focus to one and therefore mobilize the resources for another. It seems the emphasis that plays down one event and mobilizes for another might also send a message to the audience.

"In such a case, we make the decision based on strict criteria: Does the story affect masses of people? Does it change a given status quo? Does it involve more than one party or more than one country? Does it affect international politics? Is there heart-catching drama? For instance, a story like the tsunami abundantly met all these criteria. Another example of a smaller story is the 12 children who were stranded in July 2018 in a sea cave in Thailand. The story began as important tragic news but not big enough to mobilize at the beginning because if they were dead, it would have ended there. However, as it transpired, the children were alive, which added to the already existing drama and later the rescue attempts, which heightened the drama and anticipation. Therefore, the story extended for many days. When a story like this is extended and it unfolds, it lends itself to mobilization regarding time, space and teams of correspondents. Covering a story like this on an hourly basis captivates the audience," the veteran journalist pointed out.

"However, what about mobilization for just causes like the Palestinian issue?" I asked.

"I am a newsman; I go for important stories that meet the criteria regardless of the cause," he replied.

"However, what if a story includes issues with universal values or themes, such as human rights and their violations or suppression of freedom of expression? As human beings before being journalists, are we not supposed to stand up for those values?" I inquired.

"The rightness of something stands up for itself. A just cause in an important story does not need mobilizing; it will do that for itself. The Palestinian issue is about 70 years old and has always had its ups and downs. In the news cycle, many times, nothing is happening there. Therefore, you do not see it on the news. Now, when an event takes place that puts itself on the news like the Palestinian uprising in 2000, which involved the Palestinians, Israel, the US supporting Israel and

Arab countries supporting the Palestinians, that is when we mobilize, not to take sides but to report on all sides. Indeed, international law and all international resolutions are in favor of the Palestinians, which I referred to before as *the right stands for itself.* Therefore, all we have to do is to report without excluding any of the parties involved, and that is what we have always done. If one party of the conflict is not conforming with the international law and resolutions, it will consider our coverage biased. Actually, in most conflicts, there is one side that wants to reveal facts and another that continually strives to conceal them. Our job is revealing facts. Most important is to say that *being balanced* is and should be the *right cause* that we should all stand for," the seasoned journalist said.

Working in Partnership With Nongovernmental Organizations and Civil Society Groups

Unlike the mobilization model in which media outlets work primarily on their own, the *coalition* model is about working with other media outlets or civil society organizations to promote a common cause of interest to social forces. The question is: How often does Al Jazeera employ this model?

"We in the newsroom do not think in these terms. A sort of coalition model sometimes works with organizations like Medecins Sans Frontieres (Doctors Without Borders) in the case of the Saudi-led air attacks in Yemen that repeatedly killed many children who were being served by that organization in Yemen. In such cases, we can use doctors of these organizations as our sources because they are the only ones on the ground. Thanks to their good reputation, we do not coordinate with them regarding our coverage. However, there are a few cases in which we can go into coalition with other entities such as the cases that have to do with freedom of expression and protecting journalists. In the case of Al Jazeera network, such coalitions are the responsibility of the human rights and public liberties center, and once they are there, we can all work together to implement them," Negm explained.

Although this standing policy of not going directly into coalitions with other media is, for the most part, the network's policy, I had information about a few exceptions. When the network received the tapes of Usama Bin Laden's speeches through its international sources, Al Jazeera did send portions of the recordings to other international networks, including CNN, who welcomed them. Although the action was an exception to their general practice, they remain a stone's throw from the coalition model.

Mubashir: The Unfiltered World of Human Crises and Humanitarian Action

In April 2005, Al Jazeera Mubashir was launched as the first and only Arab reality TV channel, which provides a live feed 24/7 of all critical events as they happen from the North Pole to the South Pole and from the farthest east to the far west. Although US C-SPAN was the inspiration, AJM surpassed its mentor by widening the scope of its coverage to wrap around the whole world, when C-SPAN has been mostly confined within the United States.

Providing a live, unfiltered feed fit the mission of raising public awareness of the gravity of disasters, conflicts and humanitarian emergencies of all sorts. More importantly, it effectively mobilizes the public and helps to relieve distressed people through humanitarian action.

Ayman Gaballah, the first network director, believes covering humanitarian issues fits Al Jazeera's content like a hand in a glove. "Since the first day, we adopted reality TV formatting to take cameras into the lives of displaced people to live with them for days and weeks in their makeshift homes, camps and shelters. Our objective has always been to show those distressed inhumane living conditions. When we do this, we are mobilizing not only the emotion of the audiences but also humanitarian action in direct and indirect ways," Gaballah said.

"From 2016 to 2017, one of our female reporters, Hayat Al-Yamani, went to Syrian refugee camps located on the border area between Turkey and Syria and lived for weeks. She showed all the heart-wrenching details of the refugees' suffering. The year after, the same correspondent did the same with the Burmese Rohingya refugees in Bangladesh. The name of her live show was *Hayat, the Refugee*," the veteran journalist added.

However, the question is: What is the difference between what Mubashir offers the viewers and what the rest of Al Jazeera channels do?

"In traditional TV formats, you will see the tragedy in a 3–4 minute report now and then. The pictures are cut, enhanced and cleaned up, and the result usually becomes a sterilized story. In Mubashir, it is longer, more complete, unvarnished coverage," the man answered.

"We constantly receive positive feedback, but most of the time, we lack scientific field studies. Examples of positive feedback include: in the Islamic holy month of Ramadan 2018, we carried a live series called 'Stories of the Simple People' from the besieged Gaza strip. While broadcasting the series, we received many requests from average viewers to go and cover their unbelievably difficult lives. Through that live broadcast series, we managed to provide help to many of those suffering, either through private donors or through humanitarian organizations. After we finished the series, I received many calls from wealthy people asking us to turn the live series into a regular program so that they would be able to continue helping the needy," Gaballah proudly said.

"I believe that it is part of the mission of reality TV like Mubashir to play an essential role in mobilizing both individuals and states to participate more in humanitarian action. We bring humanitarian crises to viewers, not in sporadic news bulletins but around the clock, and offer our channel as a platform that allows viewers and distressed people to communicate through our broadcasts," he concluded.

Partnerships

However, Al Jazeera Mubashir did something unique that distinguished it from all other reality TV formats, which was to partner with international, regional and local humanitarian and relief organizations, including all major UN-affiliated ones. These partnerships opened doors for Al Jazeera crews, allowing reporters access to a number of disaster-stricken regions of the world. This enhanced coverage also gave the network an edge over its competitors.

"Indeed, those partnerships started during the second phase of Al Jazeera Arabic as of 2001. As we stepped into the third phase of the network, which saw the birth of Al Jazeera Mubashir in April 2005, we at AJM entered other partnerships with humanitarian agencies and relief organizations. Of course, organizing and cooperating with those agencies helped us a lot," Gaballah confirmed.

In the chapter that follows, we will now discuss Al Jazeera's third phase and the new role that journalists can play in helping relief organizations identify crises before they happen or in the early stages, in an effort to augment effective relief efforts and help those who suffer humanitarian crises in human, natural and conflict emergencies.

Notes

1. Dan Schiller. "An Historical Approach to Objectivity and Professionalism in American News Reporting," *Journal of Communication*, Volume 29, Number 4, 1979, pp. 46–57.
2. Robert McChesney, *Rich Media, Poor Democracy: Communication Politics in Dubious Times*, New York: The New Press, 2000.
3. George, Christopher, "Confession of an Investigative Reporter," *Washington Monthly*, March 1992.
4. Walter Lafeber, "The Reagan Administration and Revolutions in Central America," *Political Science Quarterly*, Volume 99, Number 1, Spring 1984, pp. 1–25.

5. David L. Protess et al., *The Journalism of Outrage: Investigative Reporting and Agenda Building in America*, New York: Guilford Press, 1992.
6. Ralph Nader, *Unsafe at Any Speed*, New York: Grossman Publishers, 1965.
7. Nellie Bly, *Ten Days in a Madhouse*, New York: Waking Lion Press, 2013.
8. Much has been written on this topic. For a comprehensive analysis, see David Corn, *The Lies of George W. Bush: Mastering the Politics of Deception*, New York: Crown Publishing, 2003. For a thorough discussion, see also "The War You Didn't See," a 2010 documentary by the award-winning British journalist John Pilger, that asks why US media were not more critical of the false claims that Iraq was developing WMD and harboring terrorists. www.globalissues.org/article/461/media-reporting-journalism-and-propagandawww.globalpolicy.org/media-and-the-project-of-empire/media-coverage-of-iraq-8-40/50436-the-war-you-dont-see-iraq-afghanistan-and-israelpalestine-.html?itemid=838
9. See Robin Andersen and Purnaka L. de Silva, *Routledge Companion the Media and Humanitarian Action*, New York: Routledge Publishing, 2017. Conclusion.
10. Clark Mollenhoff was a columnist for *The Des Moines Register*, and his book *Washington Cover-Up: How Bureaucratic Secrecy Promotes Corruption and Waste in the Federal Government*, New York: Doubleday, 1962 (2007 edition) broke new ground in investigative reporting.
11. Ryan Devereaux, "How the CIA Watched Over the Destruction of Gary Webb," *The Intercept*, September 25, 2014. https://theintercept.com/2014/09/25/managing-nightmare-cia-media-destruction-gary-webb/
12. MediaMonitor, "32 Violations and 100 Detained Journalists in Egypt in August," September 11, 2018. www.middleeastmonitor.com/20180911-media-monitor-32-violations-and-100-detained-journalists-in-egypt-in-august/
13. Beiser, Elana, "Hundreds of Journalists Jailed Globally Becomes the New Normal," *Committee To Protect Journalists*, 2018. https://cpj.org/reports/2018/12/journalists-jailed-imprisoned-turkey-china-egypt-saudi-arabia.php
14. Glenn Greenwald, Laura Poitras, Jeremy Scahill. "Welcome to *The Intercept*," February 10, 2014. https://theintercept.com/2014/02/10/welcome-intercept/

4 Case Studies and Al Jazeera's Next Phase
Protecting Journalists and Human Rights and Predicting Disaster

The Human-Centered Paradigm

Throughout its journey, Al Jazeera has always sided with human beings rather than power centers, something that has become one of the central paradigms of the network. This paradigm has translated into the unique perspective that Al Jazeera brings to its coverage of conflict. To better grasp this paradigm, we must remind ourselves of the fact that global media is either funded by states or corporations, and that is why they have always revolved editorially around one of these centers of power. They are more seriously influenced and driven by states' policies or by profit, both of which translate into power and are not necessarily in the best interests of the publics they are meant to enlighten.

Establishing Its Independence From the State

Al Jazeera's former Director General Wadah Khanfar, who led the network from 2003 to 2011, discussed how the network managed to wiggle itself out of the power-centered pattern. Indeed, Al Jazeera is officially sponsored by the country Qatar and is therefore a state-funded news outlet, not a charity. But, as Khanfar explained, "However, and for the sake of accomplishing that vision of 'human-centered paradigm,' we decided to establish a separation between the state's interests and the network's, something that necessitated very powerful field operations. We believed in magnifying field operations and minimizing the role of news agencies as the way for journalists to best interact with people on the ground, achieving authenticity and, by doing so, building and enhancing credibility." For example, well before the English-language Al Jazeera was established, the Arab-language network, AJA, established a bureau in Somalia nine months before the epic drought hit east Africa and then moved westward. The same situation occurred in other disasters both human-made and natural. Al Jazeera was on the ground reporting from the point of view of those affected by disaster. Magnifying field operations has not only achieved authenticity, and accordingly credibility, but it also developed the ability to forecast crises, the focus of this chapter, which has sharpened over time.

Early Warning Systems

How did Al Jazeera come by its ability to forecast crises, which came into being as of 2001? The network has consistently been on the ground ahead of the onset of disasters, well before most of its competitors are there. Equally importantly, crews have stayed longer, well after other news organizations have moved on to the next regional disaster. But it should not be forgotten that in many cases, the aftershocks of disasters are more severe than the first. Since a "crystal ball" is an impossibility, the answer is that a *complex early warning system* has empowered the network to fly red flags through its news coverage. This early warning system is based on a strict standard set for the network's bureau's chiefs and correspondents. These standards have been established after years of experience and the ability to recognize signals. When those indicators are properly read, they are able to predict emergencies. Also, the development of a robust network of partnerships with specialized humanitarian and human rights centers and agencies has sensitive and efficient sensors to forecast such crises. I will discuss the tools used by the network to evaluate the onset of crises now, beginning with its history of on-the-scenes journalism, and how that developed into the network's fill-blown commitment to reporting on humanitarian disasters.

"Although they contribute significantly to previewing crises concerning when and where, partnerships are not the only 'crystal balls' we have," says Ayman Gaballah, director of Al Jazeera Mubasher, channel AJM. "The practice of Al Jazeera staying behind when everyone else had gone allowed us to cover important stories that other competitors had dropped. This was the main reason for the birth of Al Jazeera Mubashir Misr channel (AJMM) right after former Egyptian President Hosni Mubarak was forced to step down from the presidency in February 2011.

"We thought at the time that things did not end just because Mubarak stepped down. We knew there was more to come. Therefore, a decision was taken to launch AJMM, which functioned until it was shut down in 2014. Our experience has developed from coverage to coverage, and so has our precision decision-making. The greater the experience we gain, the greater our ability to make an educated prediction," he explained. "As for the other components of the early warning system, there has been a mixture of first-class journalism, top-notch planning, extensive brainstorming and the encouragement of educational initiatives for journalists," Ayman Gaballah told me. Gaballah is a veteran journalist who was indicted, tried in absentia and sentenced to prison by the military in Egypt after the coup in 2013 for doing nothing more than practicing good journalism.

One of many, but little-known, examples of pushing the boundaries of innovative journalism is that Al Jazeera was the first and only international TV channel to post its flag on the North Pole. The flag posting was the brainchild of a journalist that led a deployment of a news crew on a mission to reach the North Pole. "We at Al Jazeera never turn a deaf ear or a blind eye to good ideas, nor do we exclude promising initiatives. By encouraging scientific thinking and creative initiatives, we have always managed to be a step ahead of many of our competitors and also, on many occasions, ahead of events." He went on to explain that, since the early days, and throughout the first phase of the network, between 1996 and 2001, what made the network different did not take place suddenly but actually took years of preparations. This experience and what they learned from the ways they operated was the foundation that led to the second phase of development, from 2001 to 2010.

Another important component of the complex early warning system is the diversity of the staff. The people employed by the network operation represent almost the whole world; at the forefront, the suffering South. This component of the early warning system cannot be underestimated. "Our staff has not covered events as much as they have lived them; we live our profession. As of the second phase, we strived to make Al Jazeera *a state of mind* and not just another excellent news channel. People have come from all over the world to be part of our phenomenon, many of them volunteers. Those volunteers bring their insights and foresights and have contributed greatly to the highly effective input that resulted in roadmaps through unchartered waters and unexplored issues, sometimes yet to happen," Gaballah said.

According to a senior journalist at Al Jazeera, Israel's war against the Palestinians in Gaza in 2014 is but one example of many to be foreseen by the network. In the newsroom, they could see the war coming. Therefore, several scenarios for the coverage were worked out and rehearsed even before the war broke out. That is why when Israel advanced, the newsroom could anticipate almost every move and was prepared, with crews in the right positions, holding the appropriate gear, and ready to ask the right questions. "By the way, rehearsing crisis coverage is part of our modus operandi now, and that is why and how we have been, in many cases, the primary source for news for global media," Gaballah revealed.

A Different Type of Global News Outlet

Salah Negm was one of the founders of the network even long before day one. He was the first and the most famous news director of Al Jazeera Arabic before he moved to direct the news in Al Jazeera English. Negm talked about Al Jazeera's strategy to have journalists on the ground, sometimes even before disasters happened and, moreover, to stay after all competitors had left. "We at Al Jazeera do not just cover natural disasters and wars and then leave. The consequences and the aftershocks continue for years. We follow up to report on the well-being of the people affected and whether the promises made for help after relief efforts were fulfilled. For example, when the earthquake happened in Haiti in 2010, we were the only ones who stayed for two years after the first massive tremor hit. It was part of our strategy that upholds human lives and seeks to show their sufferings. When covering disasters, our motivations for coverage were different, and that led to coverage that was also different." He explained that sensationalism and profit are not what motivates the network. "A striking volcano might be a good picture on the screen that benefits networks' bottom lines, but for us, what happened to the people there is of more importance."[1] Because continuing coverage is important to local audiences, "people know we are here to stay for them for a long time, they stay tuned to us, and that enhances our local ratings," Negm said.

A Different Mission, a Different Outcome

Negm's discussion raised another question: Since Al Jazeera's competitors seek higher ratings, then what keeps them from staying? Negm explained how Al Jazeera's business model is quite different from that of CNN. "In 2015, CNN announced the results of rating research, which showed they held 60% of the viewers in Africa. I went through the details of the research and found that the 60% were Africans whose annual income was over 60,000 Euros who represented less than 1% of the African audience. We at Al Jazeera go for the grassroots, not for the few fortunate; we go for the vast majority of people. Other networks are concerned with advertising revenue, and that is why they are after high viewerships among the few fortunate who can afford to buy. We are not. We are more focused on the masses of humans who are mostly suffering. It is just the difference between the way we at Al Jazeera calculate our viewership and how the other networks, like CNN, do theirs," Negm pointed out.

The more Al Jazeera's news director talks, the more questions are raised in my mind. He adresses a series of queries such as: Does commercializing news impact the organization's ability to forecast crises and cover them? What is the impact on its public image? Has Al Jazeera gained or lost by not commercializing the news when its competitors have done so? His answers are wide ranging. "It depends on your mission statement and whether you are out there for the wealthy minority or the underprivileged masses; profit or impact? In either case, there will be news, and it is left to the organization to decide on which direction to take. That is why the comparison between Al Jazeera and CNN and other global media is sometimes incorrect because the starting point of each of us, our operation and our interests are different. Indeed, by staying in disaster-stricken regions long after everyone is gone, Al Jazeera chooses not to commercialize news, and by doing so, we have achieved more impact. However, from a different perspective, CNN can say they achieved an edge regarding profit by moving fast from one disaster to another. It all depends on your priority and your mission. For us, profit is the audiences, impact and enlightenment," Negm explained.

Disaster Prevention

The growing sense of the need for a catastrophe prevention mission, which rarely exists within heavily commercialized media, was the main reason for the network's creation of two organizational structures within the two wings of its newsrooms: Al Jazeera Center for Studies and Al Jazeera Center for Public Liberties and Human Rights. The latter, launched in 2008, was preceded by a conference organized in Doha by the network, which most of the essential human rights organizations and humanitarian relief agencies attended.[2] The conference was concluded with the signing of Memoranda of Understanding for future cooperation between the newly established Center for Human Rights and the organizations in attendance at the conference.

Structure and Education

"The idea of the Center for Studies was to use groups of experts and analysts who are not under the daily pressure of news gathering. The job of the experts has been to engage in background research and to analyze and contextualize events for more in-depth coverage before passing this data and information to the newsroom. The focus of the Center for Human Rights is to train journalists and educate them on humanitarian law, to best enable newsrooms to tell the humanitarian side of the news. The two centers are two wings of the same network's newsrooms and do not work independently. They actively contribute in setting editorial policy at the whole network and, in many cases, in formatting and framing news," Wadah Khanfar, former director general of the network, said in an interview.

Sami Al-Haj, Al Jazeera Center of Public Liberties & Human Rights[3]

Sami Al-Haj was a cameraman shooting the US invasion of Afghanistan in 2001 for Al Jazeera, shortly before American forces "mistakenly" detained him in January 2002 according to an initial statement by the US military command in Afghanistan. However, that "mistake" did cost the man almost seven years in captivity in Guantanamo. A few months after his release on November 1, 2008, Wadah Khanfar, former director general of Al Jazeera, commissioned Al-Haj, the former Guantanamo inmate, to establish the Human Rights desk within the newsroom, which soon evolved into a unique entity within the network that no other global media has ever developed.

In long sessions of interviews for the book, Al-Haj discussed his life's most severe ordeal that resulted in that unique entity within the Al Jazeera network. "On October 15, 2001, Al Jazeera deployed me as a cameraman to cover the US invasion of Afghanistan. I left Kabul for a vacation on January 15, 2002; on my return a few days later, US troops stopped me at the checkpoint on the Pakistan-Afghanistan border," Al-Haj said. "Shortly before my detainment, the US intelligence inquired from their Pakistani counterparts about Al Jazeera's correspondent who interviewed Usama Bin Laden in September 2001. That correspondent was my colleague, Tayssir Alloni, who had already left for Qatar weeks before I first arrived in Afghanistan," Al-Haj added. "Since the Pakistanis wanted to show co-operation, and although they knew I was not the right person, they handed me over to US troops; hence I was moved to Bagram military base near Kabul," Al-Haj recalled. "Having proved to them that I first arrived in Afghanistan on October while the Bin Laden interview was done in September, an American interrogator in plain clothes laughingly told me 'obviously the Pakistanis had deceived us' and that I was going to be released," Al-Haj recounted. "However, I was detained in chains in Guantanamo for six years and seven months." Al-Haj reminisced while his face wrinkled with pain despite all the years passed. During the years of his captivity that shattered his life, Al Jazeera, along with world organizations defending freedom of journalists, spared no effort to try to gain his release.[4] While battling to end his ordeal, Al Jazeera did many interviews with heads of humanitarian and human rights organizations about Al-Haj's status. Later those efforts expanded to include the dangers threatening all journalists.

Al Jazeera's Fight for International Press Freedom

Immediately after Al-Haj's release, former director general of Al Jazeera Wadah Khanfar decided to build on the close ties that the network had woven over the years of Al-Haj's captivity and commisioned the former Guantanamo inmate to establish a desk within the newsroom to liaise between the network and human rights and humanitarian relief agencies and organizations worldwide. The desk job was to gather information, news, pictures and TV materials from those organizations about the violations against journalists all over the world and publish them through all Al Jazeera platforms. The desk, in collaboration with the newsroom, launched many campaigns to free detained and jailed journalists worldwide until they were released. Two years later, in 2010, the desk became a section within the newsroom, with TV producers assigned to handle all media material provided by the human rights organizations. "In many cases, those organizations linked us with media outlets whose journalists were abused and detained, to organize respective campaigns. We did that with many media organizations in South America and elsewhere," Al-Haj said.[5]

Eventually the network expanded the section's mission to include defending freedom of expression for all people, not only journalists and the humanitarian work in the broader sense of the term. Three years later, in 2015, the section became a department in which journalists focused on human rights issues, and humanitarian crises were added to the producers affiliated with the department whose work covered all channels of Al Jazeera, not only Arabic and English websites. Later, the department grew to a center, which has been assigned, along with its many other jobs, to train journalists on how best to deal with human rights issues legally and linguistically in addition to networking with human rights organizations and relief agencies worldwide.

By this phase, the center had its bilingual Arabic/English website in addition to specific pages on both Al Jazeera Arabic and Al Jazeera English. The center also includes three divisions, the first of which is responsible for research, treaties and partnerships with human rights organizations and humanitarian agencies, whereas the second is for monitoring human rights violations. The third division is responsible for publishing and contributing to specific TV programs on AJA and AJE and producing documentaries on how best to protect journalists against violations and ways to relieve those distressed by human-made and natural disasters. "Besides all these responsibilities, the center organizes local, regional and international events in collaboration with the UN and a variety of global organizations such as the African Union. Also, a representative office of the center was established at the UN Headquarters in Geneva to follow up on all human rights issues on an hourly basis," the former Guantanamo inmate journalist said.

Full Incorporation of Humanitarian Principles and Practices Into the Newsroom

The focus of the center is to better enable the newsroom to be a more significant part of the human side of news stories, to train journalists and to educate them in humanitarian law and legal procedures. The two centers are two wings of the network newsrooms and do not work independently of them. They actively contribute to setting the editorial policy of the whole network and, in many cases, in formatting and framing news. Journalists from the Center for Public Liberties participate in the daily editorial meetings of AJA, AJE and both the Arabic and English websites. The result has been many scoops and exclusive stories that journalists provided for all Al Jazeera platforms, in addition to establishing the appropriate contexts for many other stories. Moreover, the center has been raising red flags for many crises all over the world and, in doing so, is ensuring that Al Jazeera is the first news outlet on the scene in those crises.

Looking Through Migrant's Eyes

Many times, the center has sent some of its staff to accompany refugees on board ships forced to flee their homes in pursuit of life and freedom across the seas. The deployments were designed to tell these human-focused stories, impart the sense of the refugees' point of view of the disaster experience and to convey their catastrophic fate to the public. "We did that with famines, droughts, waves of refugees and conflicts. We at the center receive all annual reports of international human rights organizations and humanitarian relief agencies at least a month ahead of them being formally issued, so that we can be ready for broadcast and publishing in the form of coverage and campaigns," Abdulla Al-Haj, director of the center, said.

Training and Protecting Journalists

"Our editorial team operates on all of Al Jazeera's platforms and in all of the network's languages. The journalists produce original stories and content examining human rights issues around the world, complementing the network's hard-hitting news. Our safety and security team works on providing the network's staff with the best training of any media in the world, according to the UN. Training on how best to deal with hostile environments and legal protections is also essential. It is an important requirement at Al Jazeera that any staff member being deployed to a dangerous location must first receive appropriate training and safety equipment," Al-Haj added. "We at the center have issued the first world declaration to protect journalists, and we organized three world events to launch the declaration which was adopted by many international organizations, including the UN. By the numbers, Al Jazeera network has, among other international media networks, suffered the most casualties among its staff regarding the death of journalists and those jailed, detained and tortured. The reason for this is the fact that Al Jazeera's mission is to tell the truth and because the main theater of the network is the Middle East, where most of the world's conflicts are," Al-Haj pointed out.

The International Press Institute and Other Partnerships

The "campaign section" works on fulfilling one of the main objectives of the Public Liberties and Human Rights Center at Al Jazeera Media Network (AJMN)—to build strategic partnerships and cooperation protocols with human rights and humanitarian organizations across the world. This desire to join forces with key players on the international stage has led to numerous positive outcomes such as increased awareness on the part of journalists to the standards and principles of human rights and the launch of strong media campaigns to support victims of human rights violations and the promotion of the principles of justice, equality, rule of law, democracy and human rights. Also, the center began distributing International Press Institute (IPI) ID cards after officially becoming a member of the Vienna-based institute, which is one of the most prominent organizations working to protect journalists and reporters around the world. Composed of 60,000 journalists and media organizations in over 100 countries, the institute officially represents the cases of its members if they are arrested or harassed, as well as bringing their cases to the attention of relevant international bodies. It also supports and provides aid to their families.

In 2018, the Al Jazeera Media Network won the International Press Institute Chairman's Award. The award honors journalists and news outlets that "fight against great odds to ensure freer and more independent media in their country or region." When the prize was awarded, the IPI went on to say:

> Al Jazeera has been a loud voice in favor of press freedom and a transformative force on the global journalism scene since its founding 22 years ago. It continues to provide exceptional journalism despite a major effort by some in the past year to shut down the network.[6]

When asked about how far the network's visions and the objectives of the center have been translated into reality, Al-Haj said: "It is an ongoing process in which we are doing our best and have succeeded in many cases."

Partnerships and Forecasting

Although costly, significant investment in setting up mechanisms to protect not just the network's journalists but journalists worldwide and the freedom of expression of all people and their fundamental rights is both professional and moral. This robust network of relations with human rights and humanitarian organizations is part of the deep and fundamental structure of Al Jazeera, which allows the network the ability to foresee crises in the making. It goes without saying that the time of the prophets, their prophecies and their miracles has long since passed. Instead, independent and robust journalism is, and hopefully will always be, the Moses against humanitarian disasters, be they natural or human-made, including first and foremost the crisis of tyranny.

Now we will turn to a consideration of the ways in which Al Jazeera has established a style book for reporting. Language is an essential component of journalism, motivated by the desire to aide those who suffer and find resolutions to conflict. This often means establishing a new lexicon and finding news words, words that are fundamentally about healing and uniting the peoples of the world, not dividing groups, ethnicities and those with differing belief systems. These practices stand in striking contrast to the language of political division, conflict, belligerencies and war.

Notes

1. For an example of follow-up coverage of the 2010 earthquake in Haiti, see Sebastian Walker. "Haiti After the Quake," Al Jazeera, September 13, 2011. www.aljazeera.com/programmes/aljazeeracorrespondent/2011/09/201196122110280787.html

2. For the center's mission statement see, "Al Jazeera's Human Rights Mission Statement," December 9, 2012. www.aljazeera.com/humanrights/2012/11/201211s2693118958570.html
3. Al Jazeera Center of Public Liberties & Human Rights. https://ethicaljournalismnetwork.org/supporters/aljazeera-center-public-liberties-human-rights
4. Sami Al-Haj talked about his detention in Guantanamo with the *Guardian*'s press freedom reporter, Gwladys Fouche. "Sami Al-Haj, I Lived Inside Guantanamo as a Journalist," *The Guardian*, July 17, 2009. www.theguardian.com/media/2009/jul/17/sami-al-haj-Al Jazeera-guantanamo-bay-journalist
5. For an example of Al Jazeera's coverage of press freedoms, see www.aljazeera.com/news/2019/12/number-journalists-killed-2019-191217052633955.html
6. Al Jazeera, "Al Jazeera Wins the IPI's Award," June 26, 2018. https://network.aljazeera.net/pressroom/Al Jazeera-wins-ipi%E2%80%99s-award

5 The Power of Words
Between Al Jazeera's Humanitarian Stylebook and the Hateful Rhetoric of Radio Rwanda

Rwanda

"To kill a Tutsi is to book yourself a front-row seat in heaven." These words have been engraved in my mind since the heinous genocide took place in Rwanda in 1994. This type of language first came to my attention in the print media, a full four years before the genocide began, and was clearly an incitement to violence even then. As the rhetoric of hate gained momentum, anti-Tutsi articles escalated in Kangura newspapers, and they also began to publish graphic slogans such as *A cockroach cannot give birth to a butterfly*.[1] Apart from calling Tutsis cockroaches, the implied meaning is that a Tutsi cannot give birth to a Hutu or anything that is as important or as beautiful as the Hutus themselves. Strangely enough, at that time, the world stood by doing nothing to stop this type of hate speech, demonizing and dehumanizing the Tutsis, which would eventually result in a massacre. Indeed, as the situation developed and violence engulfed Rwanda, the international community would come to understand the killings as nothing less than genocide, but it would be too late. The media had declared, "The graves are not yet full."[2]

With surprising speed, the hateful rhetoric in print media evolved and migrated onto an even more influential medium, the radio.[3] Radio campaigns targeting the Tutsi minority called for their extermination in Rwanda. In the early 1990s, government-owned Radio Rwanda led the call to genocide, and by June 1993, it was joined by Radio-Television Libre des Mille Collines (RTLMC) when that station began to broadcast similar hate campaigns. It employed rowdy disc jockeys who used street language, played pop music and encouraged phone-in requests and comments. This "hip" style and engaging format was designed to appeal to the unemployed and elements of the militia. This was no accident. RTLMC was set up and financed by Hutu extremists who directed the programming mostly from the northern part of the country, where

The Power of Words 65

wealthy people in business mingled with government ministers, some even rubbing shoulders with various relatives of the president. Its backers also included the directors of two African banks and the vice-president of the infamous Interahamwe Hutu militia. From October 1993 to late 1994, Radio Rwanda was used by Hutu leaders to advance an extremist agenda with anti-Tutsi messaging and disinformation, which spread a fearful scenario that Tutsis were about to commit genocide against Hutus. Specific Tutsi targets, especially business and stores, were identified, and directives were issued on how to kill Tutsis. Those who took part in the killings were congratulated on the air.

As a roving war correspondent based in sub-Saharan Africa for eight years, I witnessed the results of such demonizing rhetoric, designed to dehumanize an entire group of people. The media stylebook of hate and lies had led to violence on an unimaginable scale. But I never thought I would see the day when Radio Rwanda–style media would be disseminated from my own country, Egypt, and many other Arab countries as well.

Al Jazeera's Evolving Stylebook Compared to Radio Rwanda–Styled Arab Media

Two years before the January 25, 2011, revolution in Egypt, I could detect the early signs of a rhetoric that mimicked the hateful language of Radio Rwanda in Egyptian media. By that time, Mubarak's regime had aged politically, and preparations were underway to pass the presidency on to the elder Mubarak son, Jamal. The recipe was old, tired and well-established. Historically, many have identified the tactics as "divide and conquer":[4] incite hatred and suspicion among the various social strata in order to control them all. This political maneuvering began by sowing discord between Christians and Muslims. As a first step, there were "terrorist" attacks on Coptic Christian churches. Then the blame was directed toward Muslim Egyptians, linking "terrorists" with Muslims generated Christians' fear, which succeeded in aligning the church with the regime. Such habitual crimes were revealed only after the revolution, when an angry public stormed State Security Headquarters. There they found documents implicating the security apparatus in these crimes. The attacks had been planned by the government, and Muslims were blamed for them, to divide Christians and Muslims.

This type of propagandistic messaging was used by private media companies owned by wealthy businessmen associated with Mubarak and his family. Another narrative based in the politics of fear can be identified as the "build and demolish" strategy. First, a scary crisis is set up, and then the regime can step up to solve the problem, usually with military or repressive force. By doing so, the regime appears to be the best option, in fact, the only option. In this way, discourse is directed away from democracy and toward authoritarianism.

One of many examples of this narrative occurred in September 2010 on a talk show on Al-Mihwar, a television channel owned by a businessman associated with Mubarak (and now with President Sisi). Two journalists appeared, one as an anchor and the other as a guest. As they spoke in frenzied terms, they began to launch a set of accusations against the Baha'i. They claimed the Baha'i were actively proselytizing by preaching their religion in southern Egypt. As unbelievable as it seemed, the two journalists named certain villages in the governorate of Sahaj, claiming most of its population had already been converted to Baha'i. The hate speech was so blatant and the incitement so intense, I feared the worst, which actually happened only a few hours later. Villagers in the area set fire to dozens of so-called "convert" houses—homes where converts to the Baha'i religion were said to be living—killing numerous innocent people in the neighborhood. Neither the direct perpetrators nor those who instigated the violence on air were ever questioned, let alone punished. In this way, religious differences and inchoate tensions were cynically exploited for political purposes. Hate speech and demonization directed toward cultural and religious subgroups leads to violence, which is frequently politically motivated in authoritarian regimes. Such political actions were taken with the objective of passing the presidency from father to son. It is important to note that these crimes were committed by a self-named "secular" regime and "liberal" private media, not by Islamists or "extremists."

After the January 25 revolution, hopes were running high for political reform, which included the possibility, indeed an expectation, that the country's press might adopt an independent stance and work toward developing more accurate journalistic practices. My hope was that the press might report events without rhetoric and bias, that I might see a journalism free of the Radio Rwanda type of "hate-to-kill" media. However, the hate-inciting tactics continued to be used by the same media tycoons and their reporters who were allied with the Supreme Council of the Armed Forces (SCAF), which took over after Mubarak stepped down on February 11, 2011. Having successfully employed the politics of fears in media messages—the "build and demolish" strategy, the perpetual building up of fear and then demolishing it through force—media repeated these news frames, but with a difference. This time the target was the public, the revolutionaries who had demanded change. The regime now accused them of being traitors and agents of the US and Zionism. At a time when the media should have adopted a skeptical stance toward the regime, they followed them down a path of destruction.

The frenzy reached its peak during the so-called transitional period, when the Supreme Council of Armed Forces choreographed a politically charged court case against the US NGOs in November 2011. In addition to demonizing the revolution, innocent people were unjustly accused and convicted, including myself. There were 42 defendants, including 14 Egyptians, and others, mostly American. The objective was to deliver a critical message to the public by way of media disinformation—*the US NGOs were behind the revolution that brought down Mubarak*. Another underlying message was that standing up to the military regime would never be allowed again. As a defendant for two years, I witnessed a trial where the media, both private and public, were instructed by the administration (the security apparatus and SCAF) and even the judiciary to defame and employ character assassination on the defendants without allowing them an opportunity to present a defense. We were defamed, criminalized and sentenced, yet no evidence was presented to the court that would justify such charges. With the intent to demonize the revolution, the military regime used hate-filled language and narratives to foment hostility, very similar to the speech designed by Radio Rwanda. Again, the objective of saving the regime justified the means, which was sacrificing innocent, respectable people and their families.

The media frenzy against the US NGOs escalated further, as fearful language and demonizing narratives were used to condemn all foreigners in the country, describing them as spies posing as tourists, human rights activists and people from other organizations, both local and foreign, including diplomats. Most TV and radio talk shows spread these lies. The media campaign resulted in unprecedented xenophobia throughout Egypt. As the regime desperately sought to hold onto power, it became time to kill, both morally and physically, and the notion of a free press was counted among the casualties.

As a journalist with years of experience covering the Middle East, I have learned that robust and independent journalism must necessarily be against all tyrant "pharaohs" and all the vices that come with them, including human-made disasters and natural ones. To do so, independent journalists should abide by long-established values, including the willingness to bring the truth to the light of day, report multiple perspectives and convey the complexity of a situation with fearlessness, fairness and accuracy. Living up to the professional standards of an independent press, which I see as the Ten Commandments of good journalism, includes carefully choosing the language of reason and accuracy. The Middle East, and the Arab world at the heart of it, did not invent journalism, which is a particularly western construction. But for generations now, a corps of Arab journalists have adopted and sought to emulate professional standards and values and strived to work and live by them. In most cases, those aspiring Arab journalists, as individuals and not as institutions, have always struggled to that end, even at the risk of their freedom, and sometimes their lives.

Exposing Injustices Using an Independent Stylebook

The classic argument for independent journalism in a democracy was set out by James Madison, a principal author of the US Constitution. Madison wrote to a friend: "Knowledge will forever govern ignorance: And a people who mean to be their governors, must arm themselves with the power which knowledge gives." It has been proven that without valid information about what leaders are *doing*, as opposed to what they *say* they are doing, there can be no real democracy. Without such information and knowledge, the powerful can more easily stifle social peace, economic progress and even the basic minimum of justice. Since it is the vehicle that carries knowledge, it is essential that media language be independent and impartial and remain unbiased when reporting on race, sect, religion, gender and ethnicity. Only then will the information be viewed as valid.

If we take Iraq as an example of many Arab countries, we will find that the political map, according to the sectarian-based quota system set by the constitution, comprises a Shiite prime minister as head of the executive, a Sunni as speaker of parliament, a Kurd as a ceremonial president and a self-autonomous government in Kurdistan. All these components of the Iraqi political map, and the media, present themselves as working toward democracy and peacebuilding and crusading against corruption. Yet in reality, we have seen historically that sectarianism and division remain the rule of the day. This is also the case in many other Arab and Middle Eastern countries.

In Iraq, and to a lesser extent Egypt, media have been using a sectarian/tribal stylebook that enhances divisions in a multireligious society. Muslims and Christian Copts are set apart from one another on the one hand, and the military is empowered across most social sectors on the other. For example, *Coptic people* is an exclusionary phrase that media have used for decades, but they should not. It is an invention of the politicized Egyptian Coptic Church. When media adopt this language, they drive a wedge into the heart of the nation. Coptic Christians make up 10% of the Egyptian population. It serves to set the religious group apart, with the implied claim that they stand alone and are not part of a greater united Egyptian society. Egyptian media simply repeated this language instead of rejecting it as the divisive phrase that it is. Such a stylebook has indeed sown the seeds of many religious conflicts in the country.

Therefore, it is imperative that media in the Arab world benefit from Al Jazeera's stylebook and adopt an independent stance, using words of inclusion that accept instead of denunciation. When difference is condemned in multiethnic and multireligious countries on the basis of race, sect, ethnicity, religion and tribe, ways of thinking about the benefits of such diversity are lost to hatred and violence. If there has been good reason historically to use a political quota system in a country like Iraq, it must not replace the universal rules of journalism. In other words, the political quota system must not be discursively reproduced, cloned to become the media modus operandi or part of the stylebook of reporting in the Middle East.

Independent Media as Society's Immune System

When people know their rulers are corrupt, they may not be able to do anything about it, and they become cynical. That is why knowledge and information is still so important. Especially in the face of such contempt for public officials, when investigative journalism reveals the truth about the powerful, other stakeholders, such as civil society organizations or honest officials, will take note of it. Social stakeholders become de facto allies for journalists working to keep the powerful accountable to the people. In this way, journalism is part of democratic practices working together to move toward social justice. This is fundamentally different from employing an emotionally charged and biased stylebook that persuades through fear and ignites social unrest.

Conversely, by bringing corruption to the light of day for all to see, journalists help protect brave individuals who confront power, and their actions and information enter the public sphere. Mark Lee Hunter, "Story Based Inquiry: A manual for Investigative Journalists," *UNESCO*, 2011, https://unesdoc.unesco.org/ark:/48223/pf0000193094/PDF/193078eng.pdf.multi. nameddest=193094. They are no longer alone and in the dark. If an opportunity later emerges for political reform, these forces may act on it. At the least, it becomes more difficult for corrupt practices to continue in the face of public exposure. The risks to wrongdoers rise; their opposition perceives their vulnerabilities; their friends are ashamed of their company. The wrongdoers may or may not care, especially if they are powerful enough to silence their enemies. However, even small streams can cut deep holes in mountain rock.

From all these perspectives, the value created by good journalism is first and foremost social and *general*. It benefits the entire society, as it benefits the media enterprise and the journalists themselves. As in the human body, strong independent media works as an immune system for society. Since it is the vehicle to convey thought and ideas, language must be free of the culture of vendetta, which has not been the case most of the time in the Arab world until a new media model was born with Al Jazeera.

The Al Jazeera Stylebook

Until Al Jazeera was born in 1996 as the first institutionalized Arab effort to instate the professional values of journalism, Arab media remained shackled in sectarianism and tribalism. Al Jazeera introduced a new set of standards for independent reporting, rejecting the destructive language of difference with regard to race, sect, religion, ethnicity and tribe that plays such a destructive role in the Middle East. Ironically, by the time efforts were underway to establish Al Jazeera, some of the western media had been undergoing noticeable changes and seemed to be moving away from independent reporting as they move closer to US military objectives. The early manifestation of these changes was evident during the CNN coverage of the First Gulf War in 1991.

There is a scene of the celebrated CNN correspondent Peter Arendt standing in his balcony room in Al-Rashid Hotel in Baghdad talking to famous anchor Bernard Shaw, and while he was looking at the sky, he said: "Are these planes ours or theirs?" Of course, Arendt did not mean CNN's warplanes; instead, he meant the planes that belonged to the US military.

I talked to Salah Negm, director of Al Jazeera English News and former director of Al Jazeera Arabic News, about the dangers of using such subjective language and whether Al Jazeera was lured by such practices and temptations. He immediately retorted: never.

"We have always named things the way they are. It is either US, Iraqi, Iranian or whatever jets they might be; there have never been possessive nouns. If you say our Qatari jets, for instance, then you are taking sides, and nobody will trust you."

"In the beginning was the word." You would find this quote not only in John but also in all divine holy books, and so in journalism. A word could be the dividing line between life and death, between enlightenment and ignorance and between right and wrong. The right diction of a media outlet can change the international media's agenda by setting the right terminology, something that Al Jazeera has done on many occasions.

The World Defined and Redefined by Al Jazeera

"Our intention has not been changing anybody's agenda. Rather, it is seeking facts, right terms and right definitions. People use language to justify their actions, sugarcoat them or shape the minds of the others in a way to support their actions or beliefs. Our job is to get to the bottom of things," Negm explained.

"When international media along with western politicians were using the word *immigrants* and sometimes *illegal immigrants* referring to Syrians escaping their country with their lives from the hell set for them by their ruler, we had to ask ourselves: What is the meaning of the term and the shades it involves? If we legally examine the plight of the majority of the fleeing Syrians, then the right term should be *refugees* and not immigrants. Immigration is a matter of choice, but in the case of Syrians, they had none. In the past, the word *immigrants* had a positive meaning in the west because it is a healthy sign of the recipient countries that improved the economy and enriched society's diversity. In the last few years, that same word was used by populist politicians in a way to indicate that those immigrants are invaders who are going to change the culture, social habits and take jobs. In the case of Syrian refugees, the west applied their false negative image on them. Therefore, we at Al Jazeera had to set the right term when referring to those Syrians to set the right context of the tragedy. Unfortunately, many international media repeat the terminology used by politicians without examining them. Here in Al Jazeera, we examine them," Negm detailed.

"More recently, in November 2018, it happened at the height of the Caravan crisis which was headed to the Mexican–US borders from Honduras.[5] I asked in the editorial meeting: What is this caravan about? Is it about sheep, horses or humans? The answer was humans. Then came the second question: What is the goal of the caravan? The answer was: Those mass of people want to cross the US borders and seek asylum. Whether legal or not, they are *asylum seekers*. So the decision was to name the caravan as it is: a caravan of asylum seekers. We did not invent a language; rather we use it to define and explain. By doing that, we stripped off all the rhetoric and cut through to the basic bottom line," the veteran journalist added.

In the summer of 2018, there were two referenda taking place at the same time: one in Catalonia, Spain, and the other in Kurdistan, Iraq. In the two referenda, the two provinces wanted to walk away from the two countries. From the outset, the two events seemed alike. It was interesting to watch how international media perceived the two events as much as it was telling about their fate. In this case, Al Jazeera was not an exception.

"Although seemingly identical, the two cases were different right from the beginning; one was secession, and the other was independence. Historically, Catalonia was an independent country until almost 150 years ago before joining Spain. That is why, when they wanted to break away from Spain, that was an attempt to seek and reclaim their independence. In the case of Kurdistan, it was never an independent state or even a non-state entity. Therefore, when they wanted to break away, this was a secession attempt. So, we named things the way they are: Catalonians seeking *independence* and Kurds seeking *secession*. We made our definition based on history and law," Negm, AJE news director, stated.

Imad Musa, head of Al Jazeera Online, made another observation about the Catalonian and the Kurdish referenda. He said that the word *independence* is so beautiful that no one can refuse to use it, as opposed to the word *secession* or *separation*, which carries a different history and connation.

"For a week before the referendum in Iraq, both AJE and the English web, being influenced by European staff, used the word *independence* for Kurds. In the meantime, Al Jazeera Arabic, AJA, since day one used the word *separation* for both cases in Iraq and Spain. Finally, AJA won, and we at the English Online followed them," Musa admitted. "On the other hand, most of the mainstream media in the west started by using the word *independence* when talking about the upcoming referendum in Catalonia because nobody ever thought about the issue from Spain's point of view: that it is a process of breaking up the state," Imad Musa also pointed out.

"The western media chose to ignore the other side of the story, that is, of the state, which had to do with the post-referenda implications; we did not, and this is the power of the word of Al Jazeera," the Arabic US-born journalist stated. "In this dilemma of the two referenda, and judging by the result, Al Jazeera withstood the pressures, which proved to be right, and its narrative won," Musa concluded. He went on to say,

> Editorial discussion about defining and redefining take place on all Al Jazeera platforms. It is an ongoing process that is more factual based than opinionated. This process is part of the routine daily editorial meetings. There are many examples of this process. One is about the several hundreds of immigration seekers that Australia has been holding on the island of Nauru, which Canberra calls a detention center. In the AJE editorial meeting, a question was asked: What does it mean a detention center? Is it a prison, a haven or a sanctuary? All answers were excluded except for the prison, and so AJE used the word that is the most accurate: a prison.

Unlike some of the international media and despite India's disgruntlement, Al Jazeera insists on using the legal expression of *line of control* referring to the dividing line between the Kashmir–Indian-controlled part and the Pakistanis'. The idea is all about sticking to the United Nations Security Council's resolutions as long as the province remains a disputed region between the two countries.

Another case: Libya's official name was the Democratic Republic of Libya, and so are many countries like the Democratic Republic of North Korea, Democratic Republic of Congo and many other dictatorial countries. Again, a question had to be answered at AJE: Are we required to spread such lies about imaginary democracies? Of course not. So, a decision was made to refer to those countries only by their names: Libya, Congo, North Korea and so forth, and by doing that, Al Jazeera was telling the truth and indirectly exposing dictatorships, all while being more accurate in its lexicon.

"When we refer to these countries by their names without the adjective *democratic*, we do not do this because we are anti them but because we are out there to tell the truth. In the meantime, when we do this, we are not seeking to make an impact on other media. Instead, we try to stick to reality and deliver facts as accurately as possible and to the best of our knowledge," Negm, AJE's director of news, confirmed.

Al Jazeera's Variable and Constant Dictionaries

Since words are the vehicle of our beliefs and thoughts that lead to definitions of right or wrong, they are a great responsibility, and so, too, is their use in the media. In this respect, Al Jazeera has always had a *constant dictionary* and a *variable one*; the latter represents the evolution process of the first. One of the driving forces behind the variable dictionary has been the coverage of disaster and conflict. The question is: Does the variable dictionary symbiotically impact the constant?

Ayman Gaballah, director of Al Jazeera Mubashir, addresses the issue of terminology this way: "Indeed, we do have a constant dictionary which includes hundreds of words and terms, mostly legally based. Examples of such legally based terms, some of which have political implications, the word *occupied* that is a legal term to describe a specific status like the occupied Palestinian and Syrian territories. Based on that legal term *occupied*, we use the term *Nakba*, which is the Arabic for the disaster when we refer to what happened to the Palestinians in 1948. In the meantime, we, from day one, decided to end that long tradition of Arab media of using words like *Majesty* or *Excellency* except when addressing kings and princes, but when addressing or interviewing presidents and ministers, we use their official titles."

When the tidal wave of Syrians fleeing the massacre, set off by Bashar Al-Assad, started crossing the Mediterranean, Al Jazeera resorted to the term used by the western media and politicians, which was *immigrants*, before changing the term to *refugees*, merely because they were made refugees. Changing the dictionary aimed at establishing the correct meaning, with its legal implications, set the media frame within a humanitarian discourse, and this and other editorial decisions at the network are examples of Al Jazeera's concern with language.

However, I wanted to understand how these changes were made and by whom. Making such linguistic changes in a situation of crisis can become a matter of life and death to millions stranded in disaster zones either human-made or natural: "I acknowledge the changing dictionary. As for who makes the changes, it is us, the journalists. In our work, every day we have new issues. What you are referring to as far as setting and resetting the right language is a process that takes place during the daily editorial meetings. For instance, the word *terrorism* has always been a subject of debate until we resorted to the *international law* that defines many legal issues like occupying the others' lands by force and so defines what the legal resistance is. According to this, we could set the language when we refer to what is terrorism and what is resistance without politicizing the issue. This process covers issues like *uprisings, revolutions* and many other history-changing events that affect the destiny of millions. As I said, these things are being raised during editorial meetings; then, after enough brainstorming, the editorial management sets or resets the language," Gaballah explained.

In the first Gulf War in 1991, we watched CNN's chief correspondent, Peter Arndt, going live from his balcony in Al-Rashid Hotel in the heart of Baghdad, talking to the principal presenter in the Atlanta headquarters, Bernard Shaw. While describing the scene, the correspondent commented on the explosions appearing on the screen, saying that they were the result of the "smart" missiles used by the US for the first time in history, claiming that those smart weapons *saved* lives.[6] 2015, 206.

During the second US invasion of Iraq in 2003, and after more than a decade after the first use of "*life-saving-smart* weapons," Al Jazeera was there and provided a completely different story about the so-called "smart weapons." Al Jazeera broke from other media coverage to report the massive "collateral damage," itself a problematic term, that was killing hundreds of innocent civilians on an hourly basis.

It is worthwhile to think about the reasons for such a dramatic, lexical shift, which took place over a 12-year period. It was the logical result of the network establishing a two-way flow of information. What had been the previous norm, the flow from North to South, was transformed by

Al Jazeera, and global news became a two-way information flow. The global south finally had a voice in world affairs.

"Of course, it was one-way street of information flow from North to South. It is an entirely different narrative, psyche and attitude. Because of its might and the Hollywood kind of TV techniques, America has prevailed and dominated most, if not all, the world media. I believe the humiliation Arabs felt when they were forced to follow a war in the heart of their world on a TV coming from CNN's headquarters in Atlanta, thousands of miles away from them, was the catalyst for the Arabs to look desperately for their own international media. This resulted in the establishment five years later of Al Jazeera. That is why between 9/11/2001 and the Iraq invasion in 2003, it was the most exciting time for me to work for Al Jazeera, which was debunking everything and questioning anything," Emad Musa, the US-born Arab journalist and head of Al Jazeera Online, told me.

Words Are Political

Language is the vehicle that carries our thoughts, and it reflects the way we perceive the world, which is not by any means static. Therefore, language is not and should not be static, either. Talking about this process of the evolving dictionary of Al Jazeera, Emad Musa went on to say, "Al Jazeera has never been a political party, nor does it have an army, but only words and journalism, different perspectives and set of values. Therefore, in the case of the influx of refugees from the Middle East, Africa and parts of Asia, Al Jazeera has always been focused on the human suffering stories of poor people that only a few care about. Usually, Al Jazeera does not make a point of what she is doing, which is almost totally based on the collective moral compass of its staff and management," he added.

"Examples of how we take a decision as far as changing our dictionary. When we noticed western media adopting the term *immigrants* and later *illegal immigrants* referring to people fleeing their countries with their lives due to conflicts started by dictatorial regimes, we had to set this right by calling them with the right word, *refugees*, like in the case of tens of thousands of Syrians fleeing hell," Musa said. "However, when the poor Syrians initially escaped the massacre set for them by Assad's regime and arrived in Turkey, western media called them *refugees*, but once they crossed Europe's borders, they immediately changed the word to *immigrants*. In Al Jazeera, we could not accept such hypocrisy, especially since we believe that words are political. Al Jazeera was the one that triggered that debate. After Al Jazeera published a blog saying why we refuse to use the word *immigrants*, as they are refugees from start to finish. That blog was synchronized on all our platforms. At that point, the western media started to be responsive. Indeed, we can wholeheartedly credit Al Jazeera for starting that kind of debate in the hallways between politicians and media: Are they immigrants or refugees?" Musa explained.

There are many examples of the ways Al Jazeera changed the dictionary of the western media. Another is the case of the language used when talking about Palestine and the Palestinians, which has long been biased thanks to the influence of both Israeli media and politicians. Imad Musa gives a stark example of one of these dramatic changes:

"During the 1950s, 1960s and 1970s, both Israeli and western media used to call Palestinian resistants *Fedayeen*, an Arabic word for freedom fighters. Later, the Israelis made a concerted effort to change that word to *saboteurs* and eventually settled on *terrorists*, and it was all a reflection of the political changes. It is a world of clashing narratives on just about all subjects and all based on words that reflect different views. It is a clash between Steve Bannon's Breitbart kind of media that calls for the inevitability of the showdown with Islam and the other liberal media that always call for peaceful coexistence. Al Jazeera has been a leading party of that conversation thanks to its credibility built over the years," Musa explained.

In 2003, during the Iraq invasion, the debate in the United States was whether TV anchors should wear am American flag pin on their suits to show patriotism. Many US journalists said that in times like these, we must give up some of our journalistic principles and show solidarity to our army and to our nation. However, by doing so, they lost their independence and made themselves part of the US Army. Although many TV anchors wore the flag pin on their suits, the whole issue was hotly debated by the American Association of Journalism. That development heralded the phenomenon of embedding journalists with the military, something that profoundly impacted many of the profession's values, let alone the language.

"At the time of the US invasion of Iraq, I was working for Al Jazeera in the Washington, DC, office. It took the American media a year and a half after the invasion to realize that they ha[d] been propagating lies such as that the Iraqis had acquired 'yellowcake' uranium and 'secret nuclear weapons.' Finally and eventually, they had to admit that Al Jazeera was right," Musa recalled. Iraq had no weapons of mass destruction. It was an invention of the intelligence services in the United States and the United Kingdom.

Judging by what happened during the Iraq invasion and the fact that they abandoned a fundamental principle of independent journalism, that is, of challenging officials, something that resulted in compromising long-standing traditions of the profession, the US media contributed to the *catastrophe* rather than contributing to a humanitarian perspective. I asked Musa if he thought that Al Jazeera, by doing something completely different, had helped indirectly to get the US media back on the right path.

"Indeed! The US media at that time behaved like any state-owned-controlled media in the third world countries by following its military, intelligence community and the defense complex blindly. They proved the saying: *If you do not stand for something, you fall for anything.* To maintain your status as the *fourth estate*, you must stick to questioning everything and challenging it. In doing so, you must stand up firm in the face of death threats such as the many we received in Al Jazeera office in DC at that time from people accusing us of being anti-America and of being a fifth column and that we should not be allowed to operate in the US. However, the US is not a monolithic society. The Iraqi catastrophe was a paramount example of everything positive and negative as far as the role of journalism is concerned," Musa explained.

One of the most dangerous aspects of language and the use of a stylebook are words that express hatred and hostility defined by race, religion, sect or even nations based on their governments' policies. Al Jazeera has been making continuous and strenuous efforts to avoid such language, to the extent that some might call it an obsession, so important is the issue for media worldwide. Even when it came to defending its own existence against the siege from Saudi Arabia, United Arab Emirates, Bahrain and Egypt, whose list of demands require shuttering the network, Al Jazeera has been very cautious with its language, to the extent of getting an independent third party to monitor its performance.

"Right from the start of the siege on June 4, 2016, we made sure that all these anti measures against Qatar and Al Jazeera did not reflect negatively on Al Jazeera speech. Accordingly, we have not reciprocated to the hate speech spewed both by the besieging country's officials and their media. Moreover, we asked the specialized international organizations that monitor hate speech in the media to watch our language and performance," Samy Al-Haj, director of Al Jazeera Center for Public Liberties, confirmed.

The Al Jazeera Study Center was directed to monitor and help formulate its lexicon to arrive at the right rhetoric, free of hate speech that characterized that of the besieging countries. Dr. Mohamed Al-Mokhtar, director of the study center, said it has been providing the channels with accurate information and facts, in-depth analyses, estimates and previewed scenarios. "The best way to avoid hate speech is to stick to the facts," he concluded.

Three-Dimensional Reporting and the World Stage

Since words and visual content on the TV always go together and must mutually connect with each other, the work of the creative department becomes an integral part of the language and the stylebook. Nawaf Al-Mansouri was the first graphic designer at Al Jazeera since its launch in 1996, and he now heads the creative department.

"The department is assigned to packaging the news in a way that is both suitable and most attractive visually, guided by what we want to deliver through our coverage and in a way that does not color the news. The task also includes dressing the presenters and newscasters regarding the colors and fashion. For instance, if we are broadcasting a program on communism, then we should be using the color red and its shades," Al-Mansouri said.

In news coverage and media campaigns, two standard models are often used by TV channels as modus operandi, their standard practices:

- *Mobilization*: This aims to influence the public by making them aware of a situation, influencing opinions and taking action for a specific cause.
- *Coalition*: This aims at establishing coalitions between a network or channel and other media organizations and civil society organizations to maximize the effect and minimize the pressures put on one organization.

The two models are most effective on the topic of human rights, freedom of expression, natural disasters and human-made disasters. The question is: Does the creative department sometimes subject its visual and audio materials to one of these models or both?

"Al Jazeera has introduced a model of its own: the *scale model*, which is about balancing differing opinions by providing a platform to all of them. Upholding this model has given Al Jazeera a higher ground both morally and professionally, and has gained the network a prestigious reputation worldwide in record time. However, Al Jazeera is biased towards values including protecting human rights and freedom of expression. The mobilization could be the part of the modus operandi of all media outlets. However, the question is: Mobilizing to what end? Is it for a just cause and human values or for just profiting politically, financially or both? The answer makes a huge difference. In Al Jazeera, we, the staff, are all mobilized and trying to get the audiences mobilized behind these human values," Al-Mansouri explained.

However, in the case of trying to mobilize the public behind just causes or human values, what are the techniques used by the creative department to have a positive effect?

"We all know that the viewer has a mind and an emotion; the colorless news addresses the mind, whereas the creative regarding visual and audio packaging addresses both the mind and emotion, or heart, and that is what we exactly do," the creative artist pointed out.

Songs are about words and music, whereas music is all about transmitting unspoken words and meanings. Al Jazeera provided unique coverage of the Egyptian revolution on January 25, 2011, during which it searched for and used certain parts of some of the old Egyptian songs and music that had long been forgotten. The old historical material had an unexpectedly profound impact on most of the Egyptians, to the extent of driving millions of them out to the streets to join their fellow revolutionaries crying out for freedom and dignity, bread, liberty and social justice. These songs and music were used in intervals between the programs and even within the news rundown. Could this be used as a case in point on how the creative wittingly can play a pivotal role in employing *artistic words* to achieve the mobilization model? Moreover, in this case, who chooses the music and songs?

"In this specific case that you raised, I was the one who chose the songs and music and the specific parts to play. However, as I said before all media all over the world are concerned with *mobilizing*. The question remains: to what end? Going back to the case in point you used, were we supposed to use in covering the Egyptian revolution Beethoven music? We are talking about an extraordinary Egyptian historical event made by all strata of the people. Therefore, all visual and audio tools should come at the same level and of the same nature. What Egypt witnessed in 2011 was a political earthquake that needed coverage at the same level. When innocent civilians come under ferocious attacks merely for demanding their fundamental rights, one cannot help but to side with them. When we decided to use songs of great Egyptian singers such as Om Kolthom, Mohamed Abdel Wahab, Abdel Halim Hafez and Sheik Imam, we used symbols that belong to all Egyptians including the army, police, and that we all Arabs are proud of. I believe what we did was compatible with the norms and values professionally, morally and politically. The decisive factor at that time was that the impact of those audio and visual tools (artistic stylebook) was in favor of freedom and democracy that unfortunately was short-lived due to the military coup in July 2013," Al-Mansouri revealed.

On June 4, 2017, Saudi Arabia, United Arab Emirates, Bahrain and Egypt besieged Qatar by land, sea and air. The four countries' list of demands was topped by the order to shutter the Al Jazeera network. How did the creative department instate the mobilization stylebook model during that life or death crisis?

"By presenting the opinions of all besieging countries in the promos made in and by the creative department. The idea was simple: if we presented all the opinions of the besieging countries, we would show how unjust, illegal and farcical that act of war is. When we presented in our promos how the besieging countries jailed and punished hundreds of their citizens for merely showing empathy to the besieged Qatar, we stripped off naked those countries. In the creative department, we have shown in the promos the contrast between the immoral low ground the besieging countries have chosen versus the high moral ground Qatar has maintained. We did the same thing in chapter heads of the news. As for that part of the siege's demand list that has been claiming the life of the network, we simply presented their demand versus our will; Al Jazeera is here to stay," Al-Mansouri defiantly concluded.

Media Self-destruction: Losing the Independent Stylebook

Undoubtedly, when media lose independence, either from political regimes, organizations or even individuals, they must pay a heavy price, especially in the case of public and private media in the Arab region. Needless to say, early symptoms that indicate a loss of independence are the use of biased, racist, sectarian, tribal and other shameful linguistic strategies of a misinformed stylebook. Many unprofessional, unethical journalists in the Arab world, who work as little more than mouthpieces for tyrannical intelligence apparatuses against their minority of independent, professional colleagues, are selfish enough not to care about the destructive results of losing their media independence. The social costs seem acceptable, as long the media organizations flourish financially. For those unethical journalists, we confirm that losing independence is the fast track to losing credibility, and the whole business of media suffers from the loss of credibility. Keeping media independence does not only serve the noble purpose by keeping democratic society healthy, but it also serves the realistic goal a assuring a profitable bottom line.

If we look at the case of *Al-Ahram*, the most prestigious and oldest Egyptian newspaper in the Middle East and Africa, over its 144 years in print, it has managed to maintain a considerable margin of independence. And this has paid off greatly with regard to profits, assets and influence. But in 2005, the son of former President Mubarak, aspiring to inherit his father's position, replaced *Al-Ahram*'s professional management with a chairman of the board and chief editor who were both willing to sacrifice the paper's independence to keep their prestigious jobs. Although *Al-Ahram* has always been known to be pro-government, it had maintained a balanced editorial policy and a relatively independent stylebook, which held a degree of skepticism toward the nation's powerful players. Soon after the management reshuffle, which was supposed to herald a new political change, that skeptical stylebook that valued accuracy and fairness was thrown out. It was replaced by a total lack of recognition for any position that opposed the consolidation of power and wealth within the Mubarak family forever. The immediate consequence of losing that relative independence was a freefall of the newspaper's daily circulation, from 1.9 million in September 2005 to 750,000 in 2007. It fell further down to half a million in 2008, and to 200,000 by 2010.[7] After the military coup of July 3, 2013, *Al-Ahram*, along with all other media, sided with the military's use of a Nazi-like dictionary that dehumanized opposition, described them as cockroaches and snakes and called for liquidating them physically. Media in Egypt, especially *Al-Ahram*, has turned this racist stylebook into an actual killing machine. As the situation deteriorated, the paper lost its long-fought credibility, and the circulation collapsed to where it stands now, at around 20,000. As for the once-successful magazines published by *Al-Ahram*, the circulation also drastically dropped, to a low of 400 copies a week, to a high of 1,500 in the "best-case scenarios." The collapse in circulation was accompanied by a freefall of the advertising revenue, which is synonymous with credibility. In 2005, the paper's net profit of advertising was 3 billion Egyptian pounds, equaling about one billion US dollars. In 2017, advertising profit dropped sharply to 120 million, equaling six million in US dollars, after the sharp devaluation of the local currency. While trying to keep up with its financial commitments, the media house had to liquidate many of its assets. As of 2016, the onetime most prestigious media conglomerate in the Middle East and Africa ended up appealing to the ruling military regime to pay the salaries of its journalists and workers. Of course, the higher editorial management kept their jobs and lucrative financial benefits by courting military support, seemingly unaware that they have been barking up the wrong tree both morally and pragmatically.

The lesson to be learned by examining the dynamics that brought down Egypt's once prestigious and influential paper is that when media give up their independence and side with military and social power centers at the expense of the public, they lose credibility, legitimacy and ultimately any influence to change the political landscape for the better. The public, the press and the hopes for democracy are all shut out of the world stage. In the worst-case scenario, when a hate-filled stylebook is employed, one that incites its audience to violence instead of informing them, it violates the very foundations of freedom of expression and the role of independent media in social history.

I still nurture the hope that Arab media will someday follow the Al Jazeera stylebook and that independent journalism can be the norm in the Middle East instead of the exception. In the next chapter, I summarize my research into best practices for professional journalism and combine those views with what I have learned over 30 years of reporting from the Middle East. I hope these professional standards will help young journalists fulfill the mandate for fair and balanced coverage, and I also hope they can help them survive in an increasingly dangerous world for reporters who tell the truth.

Notes

1. For graph propaganda covers that illustrate Kangura's dehumanizing style and discourse, see Claire Gutsmidl. "The Rwandan Genocide of 1994: 'Propaganda,'" https://hutututsi.weebly.com/propaganda.html See also Human Rights Watch. "Propaganda and Practice," www.hrw.org/legacy/reports/1999/rwanda/Geno1-3-10.htm
2. See Bill Berkeley, *The Graves Are Not Yet Full: Race, Tribe, and Power in the Heart of Africa*, New York: Basic Books, 2008.
3. See *Human Rights Watch*, "Propaganda and Practice: Media," www.hrw.org/legacy/reports/1999/rwanda/Geno1-3-10.htm
4. Tetzlaff, David, "Divide and Conquer: Popular Culture and Social Control in Late Capitalism," *Media, Culture and Society*, Volume 13, 1991, pp. 9–33, London: Sage.
5. See Robin Andersen and Adrian Bergmann, *Media, Central American Refugees and the US Border Crises*, New York: Routledge, 2018.
6. Douglas Kellner. *The Persian Gulf TV War*, Boulder, CO: Westview Press. Chapter 6, TV Goes to War.
7. As a senior editor for *Al-Ahram*, I watched these numbers fall with dismay.

6 Ethics and Values of Good Journalism in a Dictatorial Environment

Setting Professional and Ethical Standards

In an environment very hostile to press freedoms, one may think that independent media should be relatively lenient and relaxed when it comes to ethics and values in order to give itself more maneuvering space within dictatorial regimes. Well, there is nothing farther from the truth. Good independent media must not equate itself with dictatorships by giving up a high moral ground under any circumstances. In this respect, Al Jazeera, since its inception in November 1996, had to set about changing the realities of the Arab world armed with professional values and practices and principles and ethics. This chapter lays out the professional practices and principles for journalistic standards for those working in the Middle East and includes my own personal experiences and the standards developed by the network over the course of its long development. These practices draw on professional standards of journalism from many organizations, and even countries, that value freedom of speech. There are many good lists detailing codes of conduct, charters and statements made by media and professional groups outlining the principles, values and obligations of the craft of journalism.[1] The discussion here shows how these standards have been adapted to the realities of the Middle East over the course of the development of freedom of the press in the region during the later years of the twentieth century and into the twenty-first.

The ethical issues facing journalists have become far less black and white and much more grey.[2] One way for journalists to discern the ideal or normative ethical position in a given situation is to consider the risks involved. We must pose the questions: What dangers may arise for journalists who decide on a given course of action? And how can the journalists manage, reduce or eliminate those risks? The vital ethical risks that have faced

Al Jazeera over time involve three central themes: revelation, sourcing and conflicts of interest. Let us consider all three from the standpoint of risk.

First: Implications of the Revelation

In the Arab world, journalists who reveal injustice, corruption or mere incompetence are habitually accused of undermining progress, public order or simply disseminating fake news. In the worst-case scenario, they are accused of plotting to topple a regime. These have been the range of accusations against all independent journalists in the Arab world as individuals, since there is almost a total absence of an institutionalized independent press for Al Jazeera and its journalists to function within. By raising awareness of problems, according to this argument, good journalists create even worse problems.

With the Help of Others

I have been accused many times of not being patriotic and of trying to distort the image of Egypt for merely doing investigations that revealed tremendous corruption, both political and financial. One of these investigations was about a land grab by more than half of the cabinet ministers at the beginning of the 1990s. The land was huge and overlooked the Mediterranean along the northwest coast. I came to know that an accident led to the discovery of many tombs of the high-ranking generals of Cleopatra's army, led by Mark Antony, who was killed in the showdown battle with the Roman army led by Octavius. The crime was about seizing the land illegally and, moreover, destroying that historical site to avoid giving it back to the state. I published the investigation over a period of eight weeks, which resulted in the government pressing charges against me for "disseminating lies that aim to distort Egypt's reputation." The only thing that stopped the government from proceeding with the lawsuit along with their crime against cultural and human heritage was the intervention of the governments of Greece, Italy and Germany. Although I knew the dangers of going public, the crime against human heritage was too great to be ignored. My primary goal was to stop the crime from happening, and if that failed, then at least to clear my own conscience for future generations. In this example, I had to discern the normative ethics in the situation based on the risks to both letting the corrupt politicians get away with their crime or losing a precious human heritage and finally going to prison for revealing that corruption. Based on this value discernment, I had to involve other allies to stand up against a corrupt government.

Timing Is Key

However, sometimes journalists must keep silent; otherwise, their freedom or lives are going to be threatened, and the revelation will never happen. For example, when I was covering the Bosnian war at its beginning in 1992 on the Bosnian-Serb side, I was invited, along with other journalists, by the Serbian police commander in the city of Neche to inspect the site of an alleged massacre committed by the Bosnian Muslims. The whole idea was to give the impression that massacres were mutual. The site was a vast mass grave strewn with scores of dead bodies. Since Bosnian Muslims, Serbs and Croats all look alike, it was difficult to decide the identity of the victims. However, I noticed a circumcised penis showing from the ripped-off trouser of one of the victims. Since Serbs do not circumcise, it was evident that those victims were Muslims and not Serbs. I asked the police commander a direct and understandable question: Are you circumcised? "I am not a Muslim," the man answered. "So, how come this dead man is circumcised? It is obvious that the victims are Muslims, not Serbs as you claim," I confronted him. Immediately, I was arrested and tortured before I was released thanks to the human rights organizations' intervention. The point is that sometimes the instant revelation of the truth is not a wise thing to do.

Who Benefits?

Based on their experience of living under dictatorships in the Arab world, some journalists may cast doubts on the effectiveness of such revelations. Al Jazeera had the answer right from the start when it asked rhetorically: How will the problem be solved or crime stopped if no one acknowledges that it exists? If the argument continues further by asking how the situation will change if a problem is revealed, then the question should be: Who benefits most when we denounce injustice?

The answer is the victims. Whether or not something changes, victims receive comfort from knowing that they are not alone.[3] This is the minimum result of a journalistic revelation of injustice or corruption, and it fully justifies telling a story that reveals corruption, incompetence or social or governance problems.

Identify the Problems

Then a seemingly ethical/professional question may arise. Should journalists only show the problem, or should they also propose solutions? This answer is related to the value of *objectivity*, which has been thoroughly considered at Al Jazeera. It has long been thought that objective journalists do not take sides. The ethical underpinning is that journalists cannot be fair to a subject if they have a stake in the outcome. The journalists' only goal, according to this argument, is to report the facts, fully and fairly. Journalists must then allow the public to conclude the story meaning of the information and let the information lead them to solutions and change. This kind of journalism has been called the "mobilization model," one that achieves results and sets public agendas.[4] It suggests that the journalist's revelations of unknown or unappreciated facts arouse public interest and awareness and form opinions and a desire for reform. Institutional actors—regulators, prosecutors, legislators, executives—must act to satisfy that desire. In this specific case, we find, at the end of the book *All the President's Men*, Bob Woodward and Carl Bernstein[5] did not say explicitly that President Richard Nixon must resign or be impeached and thrown out of office because of the Watergate scandal; if they did, they would have been taking sides. Some of the public will think that anyway, but journalists must try to keep a distance from asserting their own proposals.

However, there is another tradition that also comes into play and offers another perspective. It is exemplified by the work of Albert Londres, the famous French reporter who is said to be the founder of investigative reporting. In 1923, he wrote a series of articles exposing the horrific conditions of the penal colony called Cayenne in French Guyana. He concluded his series for *Le Petit Parisien* with an open letter to the French government demanding specific reforms. "I am done. The government must start," he wrote.[6] His work had enormous impact, and the government decided to close the prison. This tradition goes beyond denouncing injustice to proposing changes. In the 1990s, some scholars argued that proposing solutions to contemporary problems is a core mission of media. This approach is also compatible with current developments such as the entry of large NGOs into information gathering and investigative journalism. In these instances, activist organizations are founded to work for specific change. Greenpeace, for example, does not merely tell the world that the environment has problems; it actively promotes solutions to those problems, and this defines its mission. But in this regard, we should not confuse the role of journalists with that of activists and advocates. Journalists must be transparent about their interests and objectives for a given issue and be willing to assume responsibility for the consequences of their research, for better or worse. According to the long-established tradition of Al Jazeera, objectivity is all about the approach, not the conclusion. In other words, if journalists research facts fairly and accurately and conclude by favoring a particular point of view, then they can take a side. It is mostly about the methodology of researching facts in a fair and balanced way from a position of independence.

In this respect, it is important to know that what the public wants from media is not a catalogue of information but the meaning of the information that has been catalogued. That meaning is captured in a story. In other words, reporting is not merely about finding discrete facts and disconnected information. It is about telling a coherent narrative that brings to light new perspectives and ways of thinking. That is how Al Jazeera distinguished itself regionally: by being just about the only media with a strong narrative and one of the few worldwide having a unique one.

Revelations May Hurt

Journalists frequently identify specific individuals as responsible for injustices or wrongdoing. These people may suffer consequences: They can lose their jobs or their businesses. They may face social anger or criticism. They may also be prosecuted if their wrongdoing involves crimes. The question is: Under what conditions should journalists denounce someone? What can justify the damage that we may do to someone's reputation or livelihood? Is it enough to simply not like someone? Is revenge a valid motive? To answer this question, we say that *anger at injustice is a valid source of motivation, but journalists must not work only to satisfy anger*. Journalists must look coldly at the evidence, whether they like it or not. In other words, independent journalists *do not allow anger to take them beyond what the facts will justify, and they do not leave out facts to justify their anger*. The most distinct example of vendetta-free journalism is Al Jazeera's fair and balanced coverage of its host country's total siege by Saudi Arabia, United Arab Emirates, Bahrain and Egypt. Although Al Jazeera's shuttering topped the list of the demands of the besieging countries, the network has been keen to afford those four countries' officials the opportunity to make their "case."

Subjects Have the Right to Reply

Anyone who is implicated in a story as an actor or target must be given the opportunity to reply to any questions raised by the story and to explain any facts concerning them. The only exception is if speaking with the source would result in immediate physical harm to the journalist. Applying this principle to Al Jazeera, we can easily find many examples that relate to this practice. For instance, although it was hazardous and cost the network massively, financially, politically and with journalists' lives, Al Jazeera took the risk of giving Usama Bin Laden the opportunity to reply. Meanwhile, Al Jazeera, against norms in the Arab world, was the first to afford the Israelis their right to reply. For the people in the west who may not know, such a breakthrough on the Israeli front has caused much damage to the network and its journalists.

Scoops Must Be Corroborated

From time to time, journalists are offered "scoops" that are disinformation. These can come from testimony of one or more parties in a scandal or from a supposedly confidential document. In such cases, the information should not be used unless the journalist can verify it independently.[7] Ibid, Hunter 2011. For example, years ago, several Arab newspapers published the "news" that former Israeli Minister of Foreign Affairs Tzipi Livni slept with the Palestinian Authority chief negotiator Saeb Erekat and many other Palestinian officials in return for loyalty to Israel's policies. The story was never verified. Now, *we must authenticate and verify such news*. If we fail to do so, then it is not news. Moreover, that has been the policy of Al Jazeera from the beginning.

The Prudent Use of Anonymous Sources

Journalists, particularly in the Arab world, have always been concerned about using anonymous sources. In many cases, sources have genuine reasons to request anonymity. Sometimes a source runs a risk of revenge or retaliation. In this case, it makes sense for the source to think of the journalist as representing a risk of exposure and possible retribution. Other times, a source might only seek to manipulate the journalist or damage someone else with impunity. In such a case, the anonymous source represents a high risk for the journalist's credibility and media outlet.[8] If the story is false, a mere manipulation or mistake, the journalist's career can be badly compromised and, more seriously, so can their media outlet's reputation.

According to Al Jazeera's modus operandi, journalists can corroborate an anonymous source's claims by confirming it through at least two other human sources, anonymous or not, and/or enough authenticated documents.

Whistleblowers and Insiders: Protecting the Source

Many journalists imagine that the key to good journalism, primarily investigative, is getting access to inside sources in an organization or an event. In thinking so, those journalists are being influenced by films like *The Insider*. Undoubtedly, this tactic can deliver great drama and great stories. However, it also carries significant risks, both physical and moral, some of them suffered by the source after news media lose interest in whistleblowers, who may then suffer retribution. Alternatively, they may just be forgotten as journalists move to the next story. Another risk is that the inside source is a fool or a fake. That is why it must be understood that reliance on an anonymous source creates tremendous risks. The risks include severe consequences for people who are implicated in the story and who may be blamed for things they have not done. In their book *All the President's Men*, Woodward and Bernstein admit that their mistakes cost some innocent men their jobs and reputations. Those mistakes were based on anonymous sources. At Al Jazeera, there is a code that never allows a genuine whistleblower and inside sources to be compromised at any cost. This has been an absolute rule since day one of the network.

Putting the Sources Above the Secrets

Almost every country considers the possession or release of certain official documents a crime. However, in many countries of the Arab world, the boundaries that protect official secrecy are extensive. In many cases, the rules are deliberately left unclear to allow authorities greater scope for action. Or they may even be enforced arbitrarily or illegally. A typical response of many journalists at Al Jazeera to being offered such documents, in most cases, is to accept and keep secret reports and material in a safe place. Then the reporter can ask the provider of the documents if he or she can be identified as their likely source officially. If so, the journalist must never refer to the documents. The information they contain must be documented by other means.[9]

The highly trained Al Jazeera journalists never accept confidential information without asking the source how many other people know it and if he or she can be identified as the source. Vulnerable sources can then be treated off the record, meaning the journalist cannot use the information or allude to it unless they can find another corroborating source who can verify the content and be identified officially.

Third: Absence of Conflicts of Interest and Independence Are Essential

A conflict of interest arises when a journalist has bonded with one or more actors in a given story, which may have an effect on his or her judgment and negatively influence the outcome of the story. In a region like the Arab world, where conflicts of interest are commonplace, the rule at Al Jazeera was that a violation of this code would lead to the journalist losing their job. These values and ethics could hardly be found in journalism in the Arab world pre–Al Jazeera institutionally and were only evident in a small number of independent journalists. This is one of the reasons so many countries in the region are actively working to see it shut down.

The Sociopolitical and Business Value of Good Journalism

Contrary to what many journalists think, especially in the Arab world, good journalism can serve political, social and commercial interests by providing sociopolitical actors and businesspeople with valid information able to enhance the country's immune system and its prosperity. Another way that good, independent journalism, primarily investigative, can have a significant economic impact on individuals is by helping them to avoid fraud and thieves. Good journalism reporting both on politics and economic issues will and should remain a significant genre of news reporting, mainly in underdeveloped and developing countries where respective state regulations are too relaxed or too harsh and in both cases meant to facilitate and protect corruption.

Media Independence as a Business Asset

The classic model of how good journalism creates economic value for the profession was defined by James Breiner, a leading researcher of business models for news media. In his article on news entrepreneurs titled "5 Dirty Words Journalists Have to Learn to Say Without Blushing," Breiner says:

> When I was editor of Business First of Columbus (Ohio, USA), our paper did an investigative story about a bank's behind-the-scenes manipulations to get the state to take over a failed office project. The stories scuttled the deal and caused the bank, our biggest advertiser, to cancel its contract. The reaction of Publisher Carole Williams to the cancelled contract set an example for me and gave me words to live by. The lost revenue would hurt us, she said, but would not result in cutbacks at the paper. We had other advertisers. Investigative stories strengthened our credibility and made advertisers want to be associated with us, she said. In other words, good journalism was good business. In turn, profitability safeguarded our editorial independence.[10]

This argument is not only American, as some may charge. In general, every media market supports at least one source of reliable information for each significant stakeholder group, either social, political or economic, in the society. The reason is that without this information, it is tough for various actors to conduct their affairs. Thus, while specific stakeholders may become angry with a report, others will recognize the necessity for such work on the condition that it is conducted fairly and professionally.

Now, if good journalism is good business, why do people say it is bad for media business?

The Classic Model: "Good Journalism Is Good Business"

Many publishers, editors and journalists in the Arab world and the Middle East argue that good journalism is too slow, expensive and risky for their media to undertake. Moreover, they claim that being independent of power centers, besides being risky, deprives them of benefitting financially from some tyrannical regimes. This argument emerges in discussions by naïve journalists who know little about the intricacies of the profession. If investigations are conceived, designed, carried out and archived in an efficient manner, positive financial outcomes are more likely. Some of the inefficient practices are:

- Don't waste time in the dark. An example would be when reporters compile massive amounts of information about institutions or individuals without first defining a hypothesis that would enable them to focus on viable stories.
- Poor information management creates legal risks. Reporters in a hurry often make grave mistakes, because they have not understood what sources are telling them; this happened to Woodward and Bernstein during the Watergate investigation and nearly killed their story. From another perspective, few reporters are trained in archiving techniques that could enable them to control and catalogue the information they collected. Consequently, they are only partly aware, or totally unaware, of the gaps in their documentation. They are thus exposed to damages if someone sues them.
- Opportunity costs, defined as the foregone profits from being unable to seize an opportunity, arise when journalists waste resources. For example, most of the data and documentation that reporters acquire are either never catalogued or thrown away after a project is completed. Thus, valuable information that could be repackaged for consulting or further publishing is discarded.

- Promotional strategies based on the mobilization model, rather than the coalition model, limit the branding, political and social benefits of good journalism.

Ways of Creating Value From Good Journalism: Investigation as a Case Study

Investigative reporting creates less or no value for society and journalists if it is not adequately managed, promoted and distributed. To make the best of reporting, the following principles should be observed:

- The brand is independence: Good journalism adds most and best to a media enterprise when it is perceived as independent. In other words, it does little good from a business standpoint to publish an occasional investigation if the rest of the format is corrupt or incompetent.
- Professionalism: There must be top-notch management, especially of investigations, to ensure that they are well conceived and executed. At an enterprise level, the return on investment may be measured through the increased audience or advertising revenues, new skills and capabilities, brand recognition or reputation and new information or data assets.
- Promotion: Investigations, scoops and exclusive stories must not be just thrown in. Instead, they should be announced in advance to allies who will support them. Print runs are increased for exclusive stories.
- Prepare a defense in advance: Hostile response from targets of the investigations, scoops and exclusive stories is anticipated and prepared for. Contacts are made with supporting forces in the society so that adversaries will find themselves surrounded by watchful monitors. Lawyers review material before it is published. Management in the media are fully briefed on the story and can support it.

Quality Control

The practice of fact-checking is more and more widely practiced in international media. Editors and publishers in democracies are very wary of stories that cannot pass their legal tests, and fact-checking helps ensure that the stories pass. It is unfortunate to find quality control in Arab media almost nonexistent. Al Jazeera has been religiously meticulous on quality control, more than any other international media. To that end, Al Jazeera established a central department for quality control that employs some of the best journalists, legal experts, editors, linguists and sound and visual experts to pass all the network's media products. The department staff work around the clock, 24/7.

In this respect, we must underline the essential interrelationship between quality control and code of ethics; the two are inseparable. Quality control does not only protect the journalist morally and legally, but it also ensures the audiences' right to have a high-quality product, something that enhances both the credibility and the brand of the media. Al Jazeera has succeeded in all aspects: the business value of good journalism, media independence as a business asset and ways of creating value from good journalism in a way that made it a singular model in the Arab world and the Middle East and one of the few worldwide. Arab journalists may object that the conditions needed to follow AJA model are hard to meet because of the political situation in their countries, which is indeed true. However, that situation is not eternal—it will not last forever. Having a strong role model like Al Jazeera in the region is an opportunity for change. It is important that journalists in the Arab world and the Middle East realize that if they do not hang onto each other, they will hang next to each other; solidarity is the key to survival.

Notes

1. For example, see *Ethical Journalism Network*, "Five Principles of Ethical Journalism," https://ethicaljournalismnetwork.org/who-we-are/5-principles-of-journalism
2. Mark Lee Hunter, "Story Based Inquiry: A manual for Investigative Journalists," *UNESCO*, 2011, https://unesdoc.unesco.org/ark:/48223/pf0000193094/PDF/193078eng.pdf.multi.nameddest=193094.
3. Ibid, Hunter 2011.
4. Ibid, Hunter 2011.
5. Bob Woodward and Carl Bernstein, *All the President's Men*, New York: Pocket Books, 2006. New edition, a division of Simon and Shuster.
6. Albert Londres, *Au bagne*, English translation, *In Prison*, Paris: Arléa Poche, 2008.
7. Ibid, Hunter 2011.
8. Ibid, Hunter 2011.
9. Ibid, Hunter 2011.
10. James Breiner, "5 Dirty Words Journalists Have to Learn to Say Without Blushing," *News Entrepreneurs*, July 16, 2013. http://newsentrepreneurs.blogspot.com/2011/04/5-dirty-words-journalists-have-to-learn.html

Conclusion
Global Press Freedoms Under Attack

The Trial

In August 2011, I was asked to work as a country director at the International Center of Journalists (ICFJ), a US-based organization specializing in providing training for journalists who had worked in Egypt since 2006. At the time, I was the managing editor of the prestigious Egyptian *Al-Ahram* newspaper. By November, we were setting the curriculum, when I was summoned by two investigative judges to be questioned about the work of ICFJ. One of the judges talked for hours about the "Anglo-American-Zionist" conspiracies that were targeting the territorial integrity of Egypt. No reasonable arguments could sway his determination to prosecute me. In the end, I was charged with collaborating in an "American-Zionist" plot. As it turned out, the entire case was secretly made up by military intelligence, a classic disinformation campaign. My trial began in February 2012, and over the next 19 months, the media waged a fierce campaign against me. Instead of the truth, all media published the 19-month-old indictment denying us our exculpatory evidence. I told the journalists in open court that I was the chairman of the board of one of the most prestigious media houses in the Middle East and that I would never allow my reporters to violate their code of ethics so long as I am on the board. I realized that we in Egypt had finally hit an all-time low as far as the bar of press freedom was concerned; that bar was lying flat on the ground.

An Outrageous and Unlawful Demand

Throughout my ordeal, the only media outlet willing to listen to the truth of the manufactured case was Al Jazeera, even though its crew was banned from covering the trial in the courtroom. Maybe that is why, on the list of demands of the countries attacking Qatar—Saudi Arabia,

Emirates, Bahrain and Egypt—shutting down Al Jazeera, the only beacon of freedom of expression in the Arab world, is at the top.

Despite the significant global role Al Jazeera has played as a news innovator and its long and distinguished role in disseminating humanitarian information in the Arab world, the network has faced massive attacks by autocratic and tyrannical regimes. In a push-back to freedom of the press and, importantly, freedom of expression in a region where people languished for decades in cultural prisons previously, massive attacks on the network escalated during the siege of Qatar beginning on June 4, 2017, by Saudi Arabia, the United Arab Emirates, Bahrain and Egypt. Their first demand—to shut down Al Jazeera—is a demand that has never been made previously throughout the history of sovereign nations.

In the meantime, the western states, as well as western media, have not made a forceful stand in support of Al Jazeera. But if global media do not stand up to support each other, the world will remain in the dark, without a bright light to shine on the wrongs done to humanity. Worse, the pressure has not been coming only from the tyrannical countries. Instead, over time, it has come sporadically come from western democracies as well.

The United States Military Attacks on Al Jazeera

"I remember in the lead up to the invasion, the Pentagon asked us in Al Jazeera's office in Washington, DC, to provide the exact coordinates of our office in Baghdad, and we did. The US air force bombed it, trying to kill our correspondent Tessier Alloni, whom they claimed was suspected of being a spy for Al-Qaeda. At the time he was our correspondent in Afghanistan during the invasion. Fate had saved Tayssir Alloni because, by the time the assassination order was given, he had already finished his night shift, allowing our colleague Tarek Ayoub to take over the morning shift. Instead of taking out Alloni, the US military killed Ayoub. Then, the second attempt took place when Iraqi spies followed Alloni, on his way to morning coffee with his friends at Spanish TV, to Hotel Al-Rashid in the heart of Baghdad. The Americans targeted the hotel, destroying an entire floor of the building, killing many people, including some of the Spanish TV crew.[1] A mistake made by the surveillance team of Alloni was the reason for his miraculous escape from the second attempt to assassinate him.

"Free speech needs people to defend it and sometimes sacrifice lives for it," Imad Musa, the US-born veteran journalist who heads Al Jazeera English Online, said. "I believe that the whole game now has changed.

Instead of taking journalists out, they try to discredit them even by resorting to lies. It is a bloodless new game played with lies and laws to intimidate journalists," Musa noted sadly. As much as Musa is correct about the new bloodless game of discrediting journalists in the west, the old rules still remain. The nonstop murdering of journalists by tyrannical Arab regimes proves those regimes are still using the old playbook whose chapters include murder, torture, imprisonment and even holding journalists' families as hostages, They have just added the new approach of morally assassinating them as well. Dismembering the body of Saudi journalist Jamal Khashoggi after murdering him in the monarchy's consulate in Istanbul on October 2, 2018, is evidence enough of how insanely brutal those regimes have become. In any case, the future of information is at stake. That is why we see the Arab states coming back to invest in their media—to be able to present what they want the public to perceive as "truths."

In the west, the evolution of a political system was predicated on freedom of expression. The central question now is: What are the chances of such a process taking place in the Arab world regarding an Al Jazeera–style free flow of information, leading to political evolution or revolution? If they are slim, does it mean that Arabs are destined to be trapped with the same status quo? Is it soley up to Al Jazeera, and hopefully others like it, to pave the long way to democratic change? In answering these questions, Al Jazeera journalist Imad Musa thinks the Arab world is still far away from such an evolution. But that is not all. As an Arabic US-born reporter, he has a gloomier picture of chances for political change in the Arab world. The future of media in the region looks even worse. "The Al Jazeera experiment that began in 1996, especially during its first and second phases of challenging the absence of freedom of expression in the Arab world, is not going to be repeated. I think news and information are in a process of consolidation and standardization worldwide, in general and even more so in the Arab world. What we have seen with the ascendency of Donald Trump has been the last nail in the coffin. We have stepped out of an Orwellian dystopia and into a worse one, where more and more people are asking to limit their own right to access accurate information," Musa said.

Arab Spring Revolutions From 2010 to 2013

If we assume Musa is right about the difficulty of repeating the Al Jazeera experiment in its earlier years, then it might be less challenging to re-examine the pivotal role it played in bringing the Arab Spring revolutions to the Arab world and global audiences as well. What will happen if these actions repeat themselves?

Since its inception in 1996, Al Jazeera was a virtual revolution in the Middle East and North Africa and the Arabian Gulf, where media had been controlled and heavily censored by dictatorships or autocratic rulers. Its paradigm and vision have always been to present the humanitarian side of the news. Since popular revolutions are humanitarian acts against inhuman injustice, the network entered full speed with exclusive coverage from the Tunisian uprising and Tahrir Square in Egypt, to Libya in the west, to Syria in the east and to Yemen in the southeast. Almost word for word, the world followed the Arab Spring revolutions through Al Jazeera.

"Ironically, this miserable situation we have today was the result of the free flow of information, at the beginning of the new millennium, that led to the Arab Spring Revolutions and many Occupy movements in the west. These developments unleashed a terrible response by the rulers in the Middle East to control access to information. In addition, many western countries have been taking after these dictatorships by doing the same," Musa said. "The control of access to information is killing almost all choices for today's generation of Arab journalists and future ones. What would be the choices of a 22-year-old graduate fresh from a school of journalism in any Arab country? The answer is almost none. They can join the Iranian news agency, Iranian Press TV, the Saudi TV, the Emarati, the Qatari, the Turkish or the local media outlets run mostly by Arab intelligence apparatuses. Such media are all about disinformation and propaganda, and not about information," Musa added.

State-Cotrolled Media

If we take Egypt as an example of how dictatorships control access to information by controlling media, after the military coup in July 2013 and during the five years that followed, four big media companies were set up whose main task has been to buy TV channels, websites, e-zines and newspapers. The state is now "a major force in the media and runs many TV and radio stations, websites, newspapers and magazines." In addition, in recent years, "the authorities have been increasing controls over traditional and social media to an unprecedented degree."[2] The situation became so serious that the security apparatus was setting the agenda for almost all the leading TV talk shows, radio stations and newspapers, both private and public. As the BBC states, "Many of the popular TV talk shows that once featured heated political debates have disappeared.

98   Conclusion

The charge of spreading 'false news' is widely used as a pretext to clamp down" on activists.[3]

Four companies have bought and now control the majority of media platforms in Egypt.[4] They now have a monopoly on TV dramas and cinema productions.[5] The state's control later extended to include satellite channels, many of which—including ONtv and Al Hayah—are currently controlled by EMG, owned by Eagle Capital for Financial Investment, a private equity fund founded by Egypt's General Intelligence Services (GIS).[6] One might have welcomed such a move, for it would boost both the profession and the media market. However, when we learn that both the military and General Intelligence are the owners of these four media companies, we know that media in this country has become an "active measures department," a euphemistic name usually given to departments in charge of disinformation in spy agencies, especially in former communist countries during the Cold War.[7] In other words, media in this example have become an extension of the intelligence communities.

In human bodies, the first thing viruses target is the immune system. Just as in human bodies, dictatorships first target freedom of expression and more specifically press freedom. A free press often provides immunity from the disinformation of corruption perpetrated by the state on its citizens. Dismantling the free press allows dictators to be much more successful at easily controlling the people, draining their fortunes and, more seriously, destroying their spirit. Media in many Arab countries, especially Egypt, have become lethal weapons in the hands of dictators who are killing innocent people simply because they express their opposition to oppression. Most of the so-called "terrorists" who have been the victims of extrajudicial executions were nothing but peaceful protestors. As in other countries, where despots accuse the poor of being terrorists to justify their murders, lies about being terrorists are used to justify unforgivable crimes. What do we expect from a famous Egyptian talk show host who, during his talk show, would repeatedly call the police to the studio in order to arrest those who challenged his questions? Such shameful actions revealed his close co-operation with the notorious state security apparatus. These outrageous and highly illegal arrests have resulted in the detainment of innocent people, who remain under arrest even today.

It is important to remember that Egypt was the first country to introduce journalism in the modern sense of the word to the Middle East and Africa. Egyptians became accustomed to the press by way of the French, who brought it to Egypt in 1799 during their occupation of that north African country. For decades, Egypt had been a beacon in the Arab world and the Middle East in many ways, but most notably because of its press. For this reason and others, the Egyptian case serves as a benchmark to map out the deep abyss growing throughout the Arab world and

Conclusion 99

the Middle East, into which the media and politics continue to tumble. "President al-Sisi's government has tightened its control over the internet. Hundreds of websites have been blocked and online activists have been arrested. A 2018 cyber-crime law allows the authorities to block any website deemed to threaten national security or the economy."[8] We will no longer be talking about press freedom and freedom of expression; instead, we will be talking about their absence.

A quick look at the media map in 22 Arab countries reveals that out of 558 newspapers and magazines in circulation, Egypt has the lion's share of 80, followed by Lebanon with 53, Saudi Arabia 43, Sudan 34, Palestine with 29 and Algeria with 23.[9] As for TV, there are almost 500 TV stations and channels, of which Egypt, Saudi Arabia and Qatar alone hold 128 stations. Meanwhile, there are 1,200 radio stations, of which Egypt holds the greatest number.[10] Most of these media outlets mentioned are now either controlled or owned by the intelligence apparatuses.

The fact of the matter is that Egypt now routinely violates human rights,[11] and freedom of expression is blocked with the persecution of independent journalists. International reports on press freedoms from Reporters Without Borders and the Committee to Protect Journalists (CPJ) have ranked Egypt as the country with the second-highest number of jailed journalists in the world.[12] As of December 2018, CPJ documented that 23 Egyptian journalists are imprisoned, the highest figure for Egypt since the committee began recording, while the Arabic Network for Human Rights Information (ANHRI) says the figure is as high as 63.[13] The Egyptian Journalists' Syndicate statistics indicate 32 imprisonments, including 18 cases related to journalism. However, the realistic number of imprisoned journalists amounts to 102, according to Arab Media Network. Officials at the CPJ noted a huge difference between CPJ's estimated number and the Arab Media Network's estimate, only to be told that their organization accounts only for journalists who are registered in the Egyptian Press Syndicate. This discrepancy is a consequence of Egyptian law, where TV, radio and digital journalists are not allowed to join the Press Syndicate, which is open only for print. Such shortcomings with Egyptian law do not and cannot negate the fact that the imprisoned TV and digital journalists are journalists. The real number of Egyptian imprisoned journalists puts Egypt at the top of the world, with the highest number of jailed journalists.

The "Chilling" Effect

One of the many consequences of terrorizing journalists and media is that we cease to see any journalistic challenges to the dictatorship. In December 2013, CBC, a once-private TV channel now owned by one of the companies controlled by the security apparatus, interviewed the then-minister of justice. Answering a question about the number of political prisoners, which had reached by then more than 50,000, he said: "I do not know the exact number of the detainees. The critical situation we are going through does not afford us the luxury of looking into these things now, nor can we afford to take criticism under the present circumstances." Thus spoke the minister, supposedly, of justice.

As a journalist who has worked for over a quarter of a century under a dictatorship, I am not surprised to hear such an outrageous statement from a government minister of justice. The frustrating and telling scene was to see a "journalist" sitting before the minister without challenging his outrageous statements, especially when he described justice as a luxury.

World Press Freedom Day

On May 3, 2012, I addressed an audience in Cairo, Egypt, that came together to celebrate World Press Freedom Day, an event organized by the UN's regional office in the Middle East. I opened by saying:

> On today's occasion, I wanted to start by quoting the great Martin Luther King when he said: I have a dream. However, standing trial myself for charges that authorities know for a fact are trumped up charges, makes me say: I have a nightmare.

Ironically, that celebration came at the height of my trial, in which charges had been brought for my role in attempting to train my fellow Egyptian journalists. In my worst dreams, I had never imagined being tried on intentions, let alone good ones. However, my anger on that day was not only for this reason but because I could see among the celebrating audience some of my fellow journalists whom I knew well and for many years. Yet they were following the instructions given by the executive to crucify me in their columns and coverage of the trial on a daily basis. More distressing was the fact that I could see seated among them three of the lawyers who in court proceedings requested that the judge hand down the death penalty in my case. Ironically, the same three lawyers complained to the judge in the middle of the trial that they were never given a free copy of the case papers and stated that they had to pay a fee of 800 US dollars. In other words, they demanded the death penalty for a journalist indicted in a case that they had never read, let alone studied. I was the first journalist to be forced into a cage to stand trial in a bogus case only eight months after former President Hosni Mubarak stepped down. I was handed a prison sentence with hard labor only 29 days before Field Marshal Abdel Fattah El Sisi toppled former President Mohamed Morsi on July 3, 2013. The primary purpose of the case was to defame the January 25, 2011, revolution and to make the case that it was a western plot against Egypt.[14] Whoever faked that case used state-controlled media, both public and private, to propagate that lie. These recollections are the ones that make my heart reach out to all my fellow independent journalists who stand trial all over the Arab world in the same cage that I was locked in and for all the other fellow journalists who languish in prisons now.

Closing Down Freedom of Speech and Expression

Nevertheless, rolling back press freedoms occurred not only after July 3, 2013. It began in the 1950s. However, since 1984, when I started my career as a journalist, I have never seen such repression and closing down of freedom of the press. Today, we must say "press freedom" in Egypt would be a fantastic idea, but it has nothing to do with current realities. That is why I appeal to real independent fellow journalists to hang onto each other, or else we will all hang next to each other. We need to support the last few remaining free media outlets in the Arab world, especially Al Jazeera, which leads with distinction and is a true beacon of freedom of the press.

Most Arab regimes were not satisfied with merely killing press freedom but took their drastic measures further by silencing freedom of expression among their people. Thousands went missing, others were detained and others are being erased even before having their tweets or Facebook pages erased. People from different walks of life are arrested for uttering a light criticism in private meetings or even while strolling with their friends in a mall. "Nowadays, we see Mr. Trump imitating Mr. Putin's media control. Media control is a contagious disease that we in the Arab world have been suffering from for too long. Obviously, even the US is catching this disease with Donald Trump consistently discrediting the US media that he identifies as so-called fake news. Never did I imagine I would see the day that independent journalists would have to fight, sometimes to the death, to simply propose alternative narratives and to speak freely," the Arabic US-born Al Jazeera journalist Imad Musa sadly stated.

Yet even in this situation, there seems to be a ray of hope. "If we believe there is a silver lining to any crisis, then the one of Qatar's siege has changed a long tradition in the Gulf states that has historically never opened up politically. It is providing an opportunity for Qatari officials to step up and explain the many issues to their citizens. In countries whose leaders have never been elected, they did not have to explain what they were doing for, or to, their people." Musa's argument from that perspective is that the crisis has rendered an opportunity for Al Jazeera to open up the government, and the society, to more transparency. This would be a major political shift and a game changer for the whole area. If there is a silver lining to this crisis, it will be that the next generation of young people in the region will be open to transparent discussions, able and willing to examine their differences in public forums. Such an open dialogue could reach the people and help them learn to resolve their differences civilly. "Everything is being aired now," Imad Musa concluded.

Today: The Complicated Realities

With the apparent failure of all Arab Spring revolutions and the throwback of their countries to the era of dictatorships, injustice and societal chaos at varying levels, the Arab and Muslim world has entered a complex and threatening gray phase. It is now an era where visibility is blurred, and clarity is scarce, especially concerning many vital issues of life and death, freedom and economic survival. Long-fixed paradigms are being shaken and some even shattered. Traditional alliances are falling, and seemingly awkward new ones are on the way in. Who would have thought that Israel would side with Russia on many issues in the region and that Egypt, after the 2013 change of power, would pace up to join them, along with their Arab backers and financiers? Who would have expected to see the Gulf countries backing the counter-revolution in Egypt while supporting the revolution in Syria, then taking extrajudicial steps (under the tenets of international law) to fight a war in Yemen, with horrific humanitarian consequences? Who would have expected Saudi Arabia, United Arab Emirates, Bahrain and Egypt, backed by Israel, to slam a land, air and maritime siege on Qatar? The trends nowadays in the Arab world are volatile and unpredictable. As a media organization, this is not a comfortable position to be in. This phase comes with multiple challenges, and Al Jazeera must cope with this new playing field.

In such rough and stormy seas and the changing political weather patterns in the Middle East, how do we keep our moral and professional compass pointing in the right direction? Al Jazeera Media Network broadcasts to 310 million unique homes every day. How will it accomplish the migration to a new and much more interactive media world without sacrificing its traditional type of media?

Beginning in 2010, the network planned to launch an internet-only TV channel as part of its social media strategy; however, it later became preoccupied with Arab Spring revolutions. In October 2013, it announced that Al Jazeera Media Network would establish an internet-only TV channel based entirely online, called AJ+ and based in San Francisco. In June 2014, AJMN launched AJ+. The launch was attended by Al Jazeera Media Network Director General Mustapha Souag, who stated, "AJA reshaped media in 1996 when it launched, and September 15, 2014, marks the new phase of change with AJ+." In June 2015, AJ+ became the second-largest news video producer on Facebook and the ninth-largest video producer on the platform overall. By August 2015, AJ+ released data showing that it has a 600% engagement rate on Facebook, making the network's Facebook page one of the most engaged news brands in the world. In October 2015, AJ+ announced that the channel had reached over one billion views across its platforms.

As of April 2018, its Facebook page had obtained over ten million "likes" from users. In 2018, the English version of the channel moved its operations to Washington, DC, and AJ+ ended the use of its mobile apps, moving to content distribution over its website and various social media and YouTube.

Throughout these technological, production and dissemination transformations, Al Jazeera continued to question its priorities: how to attract the younger generation below the 40-year mark by adopting newer technologies and structures without sacrificing its older audience and to keep the depth, legacy and heritage that have been built over the years. Since its launch in 1996, Al Jazeera has stayed true to its initial mission. Each new phase never negated its predecessor, and each told a humanitarian story. As a professional war correspondent in this line of journalism for almost three decades, who always believed in journalism's noble mission to stop tomorrow's corruption, criminals and tyrants from being born today, I hope and expect that Al Jazeera will continue to negotiate these headwaters successfully.

But today Al Jazeera faces a truly life-and-death challenge from the continually mounting pressures brought by the siege from Saudi Arabia, United Arab Emirates, Bahrain and Egypt, and indirectly by many others, to shutter Al Jazeera. Such a demand has never been made previously throughout the history of sovereign nations. In the meantime, the western states, as well as western media, have not taken a forceful stand in support of Al Jazeera. As the region remains embroiled in conflict, the west has not lived up to the noble principles it has long stood for, including, first and foremost, humanitarianism and freedom of the press. If global media do not stand up to support each other, the world will remain in the dark, without a bright light to shine on the wrongs done to humanity. As conflict and violence continue to spread globally, including

104 *Conclusion*

in the Middle East, global publics and humanitarian actors require more information, not less. News and information are frequently at the front lines of alleviating conflict and its human costs. Shuttering Al Jazeera and shutting down its ability to inform will only lead to more suffering.

Notes

1. Suzanne Goldberg in Baghdad, Rory McCarthy in Doha, Jonathan Steele in Amman and Brian Whitaker, "Fury at US as Attacks Kill Three Journalists," *The Guardian*, April 9, 2003. www.theguardian.com/media/2003/apr/09/pressandpublishing.Iraqandthemedia
2. BBC News, "Egypt Profile," October 23, 2018. www.bbc.com/news/world-africa-13313373
3. Ibid., BBC News 2018.
4. Media Ownership Monitor Egypt: Tamer Morsi. https://egypt.mom-rsf.org/en/owners/individual-owners/detail/owner/owner/show/tamer-morsi/
5. Mohamed al-Aswany, "An Industry Under Threat: Ramadan 2019, Brought to You by Egyptian Media Group," *Mada*, December 25, 2018. https://madamasr.com/en/2018/12/23/feature/culture/an-industry-under-threat-ramadan-2019-brought-to-you-by-egyptian-media-group/
6. Ibid., al-Aswany 2018.
7. Hossam Bahgat, "Looking into the Latest Acquisition of Egyptian Media Companies by General Intelligence," *Mada*, December 21, 2017. https://madamasr.com/en/2017/12/21/feature/politics/looking-into-the-latest-acquisition-of-egyptian-media-companies-by-general-intelligence/
8. Ibid., BBC News 2018.
9. https://arabic-media.com/arabicnews.htm Arabic Media, Arabic Newspaper Ownership
10. Ibid., Bahgat 2017.
11. Middle East Observer, "Bitter Recap: A New Report Released on Human Rights Violations in Egypt," June 30, 2016. www.middleeastobserver.org/2016/06/30/bitter-recap-a-new-report-released-on-human-rights-violations-in-egypt/
12. Elana Beiser, "China, Turkey, Saudi Arabia, Egypt are World's Worst Jailers of Journalists," *Committee to Protect Journalists*, December 11, 2019.
13. The Arabic Network for Human Rights Information, "List of the Names of Journalists and Media Workers in Egyptian Prisons Until Now," www.anhri.info/?post_type=journalist&lang=en viewed March 5, 2020.
14. Peter Osnos, "An Egyptian Journalist's Nightmare: The Ordeal of Yehia Ghanem, Who Was Convicted in Egypt's Notorious NGO Trial," *The Atlantic*, June 18, 2013. www.theatlantic.com/international/archive/2013/06/an-egyptian-journalists-nightmare/276967/

Index

"5 Dirty Words Journalists Have to Learn to Say Without Blushing" (Breiner) 90

Abu Dhabi 12
Abu Dhabi TV (television channel) 12
acquired immune deficiency syndrome (AIDS) 25–26
advertising revenues 44–45
Afghanistan 6, 19, 21, 27–28, 58–59, 95
Africa 25–27, 53, 56
AJ+ (internet-only TV channel) 103
Al-Ahram (newspaper) 6–7, 14, 18, 35–36, 80–81
Al-Ahram Center for Strategic Studies 35–36
Al-Arabiya (television channel) 12, 24
Al-Assad, Bashar 74
Al-Haj, Sami 33, 58–59, 60–61, 62, 77
Al Hayah (satellite channel) 98
Al Jazeera Arabic (AJA): Arab Spring Revolutions 96–97; attention to natural disasters 29–31; banning in Arab countries 13; breaking code of silence 15, 17–18; business model 56–57; coalition model 41–44, 47, 49, 61–62; commercialism 45; conflicts of interest 89–93; content shown on western media 28–29; corroboration 88; coverage of the second Palestinian Intifada 13; coverage of war in Afghanistan 19, 21; coverage of war in Iraq 23–24; covering humanitarian crises 6, 16, 17, 22–23, 26, 29–31, 33–34, 41–42, 45, 47, 50–51, 54; demands for shutting down 31–34; documenting history 25; early warning systems 54–55; early years 7–13, 22; ethical risks 83–93; Facebook page 103; fight for international press freedom 59–60; flag on North Pole 55; formula 23; human-centered paradigm 53; impact on international news 26–27; independence 53–54; information flow 25–27; institutional agility 29–31; language use 75–77; mission 22–23, 25, 30, 37, 56–57, 103–104; mobilization model 47–49; motto 17; newsroom of diversity and open dialogue 11–13; partnerships 3, 42, 49–51, 54, 59, 61–62; persecution of journalists 33; policy 19, 25; professionalism 11; protecting sources 88–89; quality control 92–93; ratings 45; recruitment/training of journalists 8–11; relationship between audiences and 45; as representative of marketplace of ideas 28–29; repression of journalists 46; revelation 84–88; right to reply 87–88; role in opening up public

dialogue 17–18; role of Al Jazeera Center for Studies 36–38; scale model 78–80; sharing powers and rights 42–43; siege of Qatar 31–34, 37–38, 79–80, 87, 94–95, 102; sourcing 88–89; stylebook 69–82; threat of lawsuits 45–46; training and protecting journalists 60–61; United States military attacks on 95–96; use of anonymous sources 88; use of vulnerable sources 89; variable and constant dictionaries 73–75; vendetta-free journalism 87; vision 25, 62; voice for Global South 25–26; watchdog press 38–39; world defined and redefined by 71–73
Al Jazeera Center for Human Rights 42, 59–60, 61
Al Jazeera Center for Public Liberties 33, 57, 59–60, 61, 77
Al Jazeera Center for Studies 36–38, 57–58
Al Jazeera English (AJE) 8, 10, 18, 22, 30, 53, 56, 70
Al Jazeera English (Philip) 8
Al Jazeera English Online 12, 27, 72–73
Al Jazeera Media Network (AJMN) 61, 102–103
Al Jazeera Mubasher (AJM) 10, 24, 50–51, 54, 73
Alloni, Tayssir 58
All the President's Men (Woodward and Bernstein) 86, 89
Al-Mansouri, Nawaf 78
Al-Mihwar (television channel) 66
Al-Mokhtar, Mohamed 36–38, 77
Al-Yamani, Hayat 50
American Association of Journalism 76
American Broadcasting Company (ABC) 27
anonymous sources 88
anti-war movement 28–29
Antony, Mark 84
Arabic Network for Human Rights Information (ANHRI) 99
Arab Spring Revolutions 96–97

Arendt, Peter 70, 74
Asia 25

backlash 44–46
Baha'i 66
Bahrain 31, 33, 37, 79–80, 87, 95, 102
Bangladesh 50
Bannon, Steve 76
benefits 85
Bernstein, Carl 86, 89
Bin Laden, Usama 28, 58, 87
Bly, Nellie 40–41
Bouteflika, Abdel Aziz 13
Breiner, James 90
Breitbart News 76
British Broadcasting Corporation (BBC) 8, 9, 10, 23, 27, 30, 97
Burmese Rohingya refugees 50
Bush, George W. 41

Camp David peace negotiations 19
Catalonian referenda 72
Cayenne (penal colony) 86
CBC (private TV channel) 99–100
censorship 13
classic model 91–92
Cleopatra 84
CNN (television channel) 23, 24, 27, 30, 56–57, 70, 74–75
coalition model 41–44, 47, 49, 62, 78, 92
code of ethics 16–17
code of silence 15, 17–18
Columbia Broadcasting System (CBS) 27
commercialism 45
Committee to Protect Journalists (CPJ) 99
Commonwealth Summit 25–26, 37
Confessions of an Investigative Reporter (George) 39
conflicts of interest 89–93
Congo 73
corroboration 88
corruption 44, 84
C-SPAN (cable network) 10, 50

dictatorships 12, 16
dictionaries 73–75
disaster prevention 57

Eagle Capital for Financial Investment 98
early warning systems 54–55
East Africa 29
Egypt: *Al-Ahram* 6–7, 14, 18, 35–36, 80–81; "Anglo-American-Zionist" conspiracies 92–93; journalists from 8; media frenzy against US NGOs 66–67; parliamentary elections 14–16; repression of journalists 55; resignation of Mubarak 54; revolution in 65, 79; sectarian/tribal media stylebook 68–69; siege of Qatar 31–34, 37–38, 79–80, 87, 95, 102; Six Days War 35; state media 98–99; story on seizure of historical site 44, 84
Egyptian Journalists' Syndicate 99
Egyptian TV 10
Elizabeth II, Queen of England 25–26, 37
Erekat, Saeb 88
ethical risks: conflicts of interest 89–93; revelation 84–88; sourcing 88–89
Europe 25

Facebook 103
Fake News 41
First Persian Gulf War 23
freedom of expression: anti-war movement 28; coalition model 49; impact of Al Jazeera on 12, 15–16, 22, 33, 59, 78, 82, 95; oppression in Egypt 14; relations with human rights and humanitarian organizations 62; repression of 46–47, 48, 96, 98–99, 101; training journalists to perceive 8–9

Gaballah, Ayman 10–11, 24–25, 31, 50–51, 54–55, 73–74
Gaza 21, 24, 27, 50–51, 55
George, Christopher 39
Germany 84
Global South 2, 18, 20, 23, 25, 26, 37, 75
good journalism 90
Greece 44, 84

Greenpeace 86
Groppi Café 18–19
Guantanamo Bay detention camp 58

Haiti 30, 56
"hate-to-kill" media 65–66
Hayat, the Refugee (TV show) 50
Heikal, Mohamed Hasnain 35
"Historical Approach to Objectivity and Professionalism in American News Reporting, An" (Schiller) 38
human-centered paradigm 53
human immunodeficiency virus (HIV) 25–26
humanitarian crises 6, 16, 17, 22–23, 26, 29–31, 33–34, 41–42, 45, 47, 50–51, 54
humanitarian discourse 74, 77, 95, 97, 103–104
humanitarian law 57, 60
humanitarian organizations 57, 58–62
Hutu 65

independent journalism 80–82, 90–91
India 73
information flow 25–27
Insider, The (film) 89
inside sources 88–89
Intercept, The (online news source) 46–47
International Center for Journalists (ICFJ) 46, 94
International Press Institute (IPI) 61–62
investigative reporting 91–92
Iran 30, 97
Iran–Contra scandal 43
Iranian Press TV 97
Iraq 8, 21, 23, 27, 41, 72, 76
Israel 30, 35, 48, 55, 88
Italy 44, 84

Journalism of Outrage, The (Protess) 40

Khanfar, Wadah 22, 27, 42, 45, 53, 57–58, 59
Khashoggi, Jamal 96

Korea 30
Kurdish referenda 72

language 75–77
Latin America 25
lawsuits 45–46
Lebanon 21, 27, 99
Libya 73
Livni, Tzipi 88
Londres, Albert 86

Madison, James 68
Malaysia 27
"Man of The Year" (award) 6–7
marketplace of ideas 28–29
MBC Group 20
McBride Report 26
Middle East Monitor 46
misinformation 41
mobilization model 40–41, 47–49, 78, 86, 92
Mollenhoff, Clark 43
Mubarak, Gamal 14, 65, 80–81
Mubarak, Hosni 14–15, 54, 65–66
Musa, Imad 12–13, 17–18, 19, 27–29, 32, 72–73, 75–77, 95–97, 101–102

Nader, Ralph 40
Nasser, Gamal Abdel 35
National Broadcasting Company (NBC) 27
natural disasters 29–31
Negm, Salah 8–10, 22–23, 26–27, 30–31, 32–33, 47–48, 49, 56–57, 70–72
Nixon, Richard 40
non-governmental organizations (NGOs) 41, 43–44, 46, 67, 86
North Korea 30, 73
North Pole 55

objectivity 85–87
ONtv (satellite channel) 98
Operation Desert Fox 23
Operation Desert Storm 23
Organization for Security and Co-operation in Europe (OSCE) 14

Pakistan 27, 58, 73
Palestine 99
Palestinian Authority 88

Palestinian Intifada 13, 48–49, 55
partnerships 42, 49–51, 54, 59, 61–62
Petit Parisien, Le (newspaper) 86
Philip, Seib 8
Protess, David L. 40
Putin, Vladimir 46

Qatar 31–34, 37–38, 53, 77, 79–80, 94–95, 99, 101–102
quality control 92–93

Radio Netherlands 8
Radio Rwanda 64–67
Radio-Television Libre des Mille Collines (RTLMC) 64
ratings 45
reality TV 50–51
refugees 50, 60
relationship between audiences 45
relief organizations 6, 41–42, 51, 57, 58–62
Reporters Without Borders 99
repression 46–47, 48, 96, 98–99, 101
Reuters (news organization) 26
revelation 84–88
right to reply 87–88
Russia 46
Rwanda 64

San Jose Mercury News (newspaper) 43
Saudi Arabia 31, 33, 37, 79–80, 87, 94, 97, 99, 102
scale model 78–80
Schiller, Dan 38
"scoops" 88
Shaw, Bernard 70, 74
Six Days War 19, 35
Somalia 29, 53
Souag, Mustapha 103
sourcing 88–89
South Africa 25
Spain 72
stakeholders 39–40, 69
state media 8, 16, 18, 76–77, 97–99
'Stories of the Simple People' (TV series) 50–51
stylebook: Al Jazeera 69–82; sectarian/tribal media stylebook 68–69
Sudan 8, 99

Supreme Council of the Armed Forces (SCAF) 66–67
Syria 50, 97
Syrian refugee camps 50

Ten Days in a Madhouse (Bly) 40–41
Thailand 48
timing 85
Trump, Donald 46, 101
Turkey 50
Tutsi 64–65

Uganda 25, 37
United Arab Emirates 31, 33, 37, 79–80, 87, 94–95, 102
United Kingdom (UK) 23
United Nations (UN) 29, 60, 73
United Nations High Commission for Refugees (UNHCR) 42
United States (US): anti-war movement 28–29; bombing of Iraq 23; court case against NGOs in Egypt 66–67; C-SPAN 10, 50; First Persian Gulf War 23, 70, 74; influence of Al Jazeera on AIDS media coverage in Africa 26; international news divisions based in 2; invasion of Iraq 21, 28, 41, 76; legal action against journalists and media outlets 45–46; media as "the enemy of people" 46; military attacks on Al Jazeera 95–96; mobilization model 40–41; National Security Agency 46–47; Operation Desert Fox 23; Operation Desert Storm 23; US–North Korea summit 30; war in Afghanistan 21, 27–28, 58–59, 95; watchdog press 38–39
Unsafe at any Speed (Nader) 40

vendetta-free journalism 87
vulnerable sources 89

watchdog press 38–39
Webb, Gary 43
Weisman, Ezer 18
western media 2, 5, 12, 17–18, 23, 26–28, 30, 70, 72, 74, 75–76, 95, 103
whistleblowers 88–89
Woodward, Bob 86, 89
World Press Freedom Day 100–101

Zimbabwe 27

For Product Safety Concerns and Information please contact our EU representative GPSR@taylorandfrancis.com
Taylor & Francis Verlag GmbH, Kaufingerstraße 24, 80331 München, Germany

www.ingramcontent.com/pod-product-compliance
Lightning Source LLC
Chambersburg PA
CBHW051754230426
43670CB00012B/2285